£24. 00
9/88

W £18. 00

THE
NUMERICAL
SOLUTION OF
ORDINARY
AND PARTIAL
DIFFERENTIAL
EQUATIONS

THE NUMERICAL SOLUTION OF ORDINARY AND PARTIAL DIFFERENTIAL EQUATIONS

GRANVILLE SEWELL

Mathematics Department
University of Texas at El Paso
El Paso, Texas

ACADEMIC PRESS, INC.

Harcourt Brace Jovanovich, Publishers

Boston   San Diego   New York
Berkeley   London   Sydney
Tokyo   Toronto

ACADEMIC PRESS, INC.
1250 Sixth Avenue, San Diego, CA 92101

United Kingdom Edition published by
ACADEMIC PRESS INC. (LONDON) LTD.
24–28 Oval Road, London NW1 7DX

Library of Congress Cataloging-in-Publication Data
Sewell, Granville.
   The numerical solution of ordinary and partial differential
equations / Granville Sewell.
      p.   cm.
   Bibliography: p.
   Includes index.
   ISBN 0-12-637475-9
   1. Differential equations—Numerical solutions—Data processing.
2. Differential equations. Partial—Numerical solutions—Data
processing.   I. Title.
QA372.S4148   1988
515.3'5—dc 19   87-28915
                                                                      CIP

Printed in the United States of America
88 89 90 91   9 8 7 6 5 4 3 2 1

In memory of my father,

    Edward G. Sewell (1919 – 1987)

who made us understand the words: "whoever would be great among you must be your servant."

Table of Contents

Preface

The roots of this text can be traced back to a short syllabus entitled "Solución Numérica de Ecuaciones Diferenciales," written in 1974 for a course at Universidad Simon Bolivar in Caracas. It is still a short book to have as wide a scope as implied by the title, but we believe it gives a solid introduction to the computer solution of ordinary and partial differential equations, appropriate for a senior level undergraduate or first year graduate course.

The student taking a course based on this text should have had introductory courses in multivariate calculus, linear algebra, and general numerical analysis. A formal course in ordinary or partial differential equations would be useful but is not essential, provided the student has been exposed to such equations in applications or in the numerical analysis course.

After a review of direct methods for the solution of linear systems in Chapter 0, with emphasis on the special features of the linear systems that arise when differential equations are solved, the following four chapters introduce and analyze the more commonly used finite difference methods for solving a variety of problems, including both ordinary differential equations (ODEs) and partial differential equations (PDEs), and both initial value and boundary value problems. The techniques studied in these chapters are quite easy to implement, and after finishing Chapter 4 the student should be able to solve a wide range of differential equations. The finite difference methods used to solve partial differential equations in Chapters 2–4 are mostly classical low order formulas, easy to program but not ideal for problems with poorly behaved solutions or (especially) for problems in irregular multi-dimensional regions. More complicated finite difference techniques are not studied in this text. It is our philosophy that finite difference methods are still useful only because they are easy to program and to understand. When these methods become complex (e.g., when problems with irregular domains are solved), they cease to be attractive, and finite element methods should be used.

Chapter 5 contains an overview of the basic ideas behind the finite element method. After finishing this chapter, the student should have a good idea of what the finite element method is all about, and should even be able to use it to solve many simple problems. However, the student who wants to be able to write programs that efficiently solve more difficult problems should continue his/her study using the text *Analysis of a Finite Element Method: PDE/PROTRAN*, by Granville Sewell [Springer-Verlag, 1985]. This is a reference book for PDE/PROTRAN, IMSL's partial differential equation package (see Section 5.6), but it can be used as a supplementary text for a course in the numerical solution of differential equations. Chapter 5 provides an excellent introduction for the PDE/PROTRAN text, since it examines many of the ideas discussed in that book in a simpler, less general, context. The PDE/PROTRAN book can be considered as a companion text for this one, and a course that covers both would provide a good practical and theoretical understanding of the numerical solution of differential equations.

An important feature of the current text is that FORTRAN77 programs are given that implement many of the methods studied, and the reader can see how these techniques are implemented efficiently. Machine-readable copies of the FORTRAN77 programs displayed in this book are available upon request from the author.

0

Direct Solution of Linear Systems

0.0 Introduction

The problem of solving a system of N simultaneous linear equations in N unknowns arises frequently in the study of the numerical solution of differential equations: when an implicit method is used to solve an initial value problem, and when almost any method is used to solve a boundary value problem. Direct methods to solve such systems are based on Gaussian elimination, a process whose basic ideas should be familiar to the student. However, those systems that arise during the numerical solution of differential equations tend to have certain characteristics that can be exploited using appropriate variations to the basic elimination algorithm. In this chapter we review Gaussian elimination and look at some of these special characteristics.

0.1 General Linear Systems

The basic idea behind Gaussian elimination is to reduce the linear system $A\mathbf{x} = \mathbf{b}$ to an equivalent triangular system by interchanging rows (equations) and adding a multiple of one row (equation) to another. Then the equations of the reduced (triangular) system are solved in reverse order, by back substitution. The reduction is done systematically, zeroing all the subdiagonal elements in the first column, then those below α_{22}, etc., until all subdiagonal elements are zero and the system is triangular. To zero the subdiagonal element $\alpha_{ik}(i > k)$, we add $-\alpha_{ik}/\alpha_{kk}$ times the kth row to the ith row (Figure 0.1.1) and add the same multiple of b_k to b_i. If α_{kk} (called the

$$
\begin{array}{cccccc|c}
\times & \times & \times & \times & \times & \times & \times \\
0 & \times & \times & \times & \times & \times & \times \\
0 & 0 & \alpha_{kk} & \times & \times & \times & b_k \\
0 & 0 & \times & \times & \times & \times & \times \\
0 & 0 & \alpha_{ik} & \times & \times & \times & b_i \\
0 & 0 & \times & \times & \times & \times & \times
\end{array}
$$

Figure 0.1.1

"pivot") is zero, of course, we must switch row k with another row that contains a nonzero element in the kth column. We do not want to switch with a row above the kth row, because that would cause some of the elements in previous columns to become nonzero again, after they have been zeroed. Thus we must select one of the elements $\alpha_{kk}, \alpha_{k+1,k}, \ldots, \alpha_{Nk}$ as the next pivot and bring it to the pivot position by switching its row with the kth row. If all of these potential pivots are also zero, the matrix is singular, and we must give up. To see this, notice that in this case the last $N - k + 1$ rows of A each contain nonzero elements only in the last $N - k$ columns. These rows must be linearly dependent, because a maximum of $N - k$ independent rows can exist in an $N - k$ dimensional space. Since Gaussian elimination does not alter the rank of a matrix, the original matrix must have had dependent rows, and therefore it must have been singular.

On the other hand, if the potential pivots are not all zero, we face the question of which to use. If pivoting is only done when $\alpha_{kk} = 0$, this can lead to bad numerical results, for a nearly zero pivot is almost as bad as a zero one, as the following example illustrates.

Consider the two by two linear system

$$
\begin{bmatrix} \varepsilon & 1 \\ 1 & 1 \end{bmatrix} \begin{bmatrix} x \\ y \end{bmatrix} = \begin{bmatrix} 1 + \varepsilon \\ 2 \end{bmatrix},
$$

which has $(1, 1)$ as its exact solution. If ε is very small, this system is *not* ill-conditioned (nearly singular). Its determinant is $\varepsilon - 1$, or almost -1. Yet when it was solved on an IBM 4341 (single precision arithmetic) using Gaussian elimination without pivoting, bad results were obtained when ε was small. With $\varepsilon = 10^{-7}$ the solution (1.192, 1.000) was calculated and with $\varepsilon = 10^{-8}$ the solution (0.0, 1.0) was found. With rows 1 and 2 switched, the results were accurate to machine precision for ε arbitrarily small.

Thus a possible pivoting strategy is to pivot when $|\alpha_{kk}|$ is smaller than some threshold. But how small is too small? It is much easier, and safer, to simply choose as the pivot the potential pivot which is largest in absolute value. This is called "partial pivoting" and although it is not 100% safe (if the first equation in the above system is multiplied by $2/\varepsilon$, the partial pivoting strategy would require no pivoting, but this would still be as disastrous as before), it is what is used by almost all sophisticated linear equation solvers, and, in practice, it almost never gives rise to serious roundoff errors when a well-conditioned system is solved.

Figure 0.1.2 shows a simple FORTRAN77 implementation of Gaussian elimination with partial pivoting.

The amount of work required to solve a linear system when N is large can be estimated by looking at this program. Careful study shows that for large N the DO loop 30, where a multiple of row K is added to row I, dominates the CPU time. Compared to loop 30, all other portions of the code require negligible time, including the row switching and the back substitution (note that loop 30 is the only triple-nested loop in the program). Most of the work is done, then, by

```
      DO 50 K=1,N-1
          DO 40 I=K+1,N
              DO 30 J=K,N
                  A(I,J) = A(I,J)+AMUL*A(K,J)
30            CONTINUE
40        CONTINUE
50 CONTINUE
```

The number of times the statement $A(I, J) = \ldots$ is executed then is

$$\sum_{K=1}^{N-1} \sum_{I=K+1}^{N} \sum_{J=K}^{N} 1 = \sum_{K=1}^{N-1} (N - K)(N - K + 1).$$

Making the change of variable $L = N - K$, this becomes

$$\sum_{L=1}^{N-1} L(L + 1) \approx \sum_{L=1}^{N} L^2 = \frac{N(N + 1)(2N + 1)}{6}.$$

```
      SUBROUTINE LINEQ(A,X,B,IA,N)
      IMPLICIT DOUBLE PRECISION(A-H,O-Z)
C
C     ARGUMENT DESCRIPTIONS
C
C     A -(INPUT) A IS THE N BY N MATRIX IN THE LINEAR SYSTEM A*X=B,
C        ACTUALLY DIMENSIONED IA BY IA IN THE CALLING PROGRAM.
C        (OUTPUT) A IS DESTROYED.
C     X -(OUTPUT) X IS THE SOLUTION VECTOR OF LENGTH N.
C     B -(INPUT) B IS THE RIGHT HAND SIDE VECTOR OF LENGTH N.
C     IA-(INPUT) IA IS THE FIRST DIMENSION OF MATRIX A, AS ACTUALLY
C        DIMENSIONED IN THE CALLING PROGRAM (IA.GE.N).
C     N -(INPUT) N IS THE NUMBER OF EQUATIONS AND NUMBER OF UNKNOWNS
C        IN THE LINEAR SYSTEM.
C
      DIMENSION A(IA,*),X(N),B(N)
C                             COPY B ONTO X
      DO 5 I=1,N
    5 X(I) = B(I)
C                             BEGIN FORWARD ELIMINATION
      DO 50 K=1, N-1
         IBIG = K
         BIG = ABS(A(K,K))
C                                FIND THE LARGEST POTENTIAL PIVOT
         DO 10 I=K,N
            IF (ABS(A(I,K)).GT.BIG) THEN
               BIG = ABS(A(I,K))
               IBIG = I
            ENDIF
   10    CONTINUE
         IF (BIG.EQ.0.D0) GO TO 80
C                                SWITCH ROW K WITH THE ROW (IBIG) CONTAINING
C                                THE LARGEST POTENTIAL PIVOT
         DO 20 J=K,N
            TEMP = A(IBIG,J)
            A(IBIG,J) = A(K,J)
            A(K,J) = TEMP
   20    CONTINUE
         TEMP = X(IBIG)
         X(IBIG) = X(K)
         X(K) = TEMP
         DO 40 I=K+1,N
            AMUL = -A(I,K)/A(K,K)
            IF (AMUL.EQ.0.D0) GO TO 40
C                          ADD AMUL TIMES ROW K TO ROW I
            DO 30 J=K,N
               A(I,J) = A(I,J)+AMUL*A(K,J)
   30       CONTINUE
            X(I) = X(I)+AMUL*X(K)
   40    CONTINUE
```

```
      50 CONTINUE
         IF (A(N,N).EQ.0.DO) GO TO 80
C                            BEGIN BACK SUBSTITUTION
         X(N) = X(N)/A(N,N)
         DO 70 K=N-1,1,-1
           DO 60 J=K+1,N
             X(K) = X(K)-A(K,J)*X(J)
      60     CONTINUE
           X(K) = X(K)/A(K,K)
      70 CONTINUE
         RETURN
C                            IF THE LARGEST POTENTIAL PIVOT IS ZERO,
C                            THE MATRIX IS SINGULAR.
      80 PRINT 90
      90 FORMAT (23H THE MATRIX IS SINGULAR)
         RETURN
         END
```

Figure 0.1.2
Gaussian Elimination with Partial Pivoting

This last equality can be found in any mathematics handbook and is proven by induction. Ignoring lower order terms in N, then, the total number of times this statement is executed (and thus the total number of multiplications done), is $N^3/3$.

0.2   Systems Requiring No Pivoting

The linear systems encountered while solving differential equations numerically often have relatively large diagonal elements, and pivoting is sometimes unnecessary. One class of systems for which pivoting is unnecessary is characterized in the following definition:

Definition   A matrix $[\alpha_{ij}]$ is called *diagonal-dominant* if

$$|\alpha_{ii}| > \sum_{j \neq i} |\alpha_{ij}| \qquad \text{for } 1 \leq i \leq N.$$

The matrix is said to be "weakly" diagonal-dominant if the "greater than" in the above inequality is replaced by "greater than or equal to."

It will now be shown that

Theorem 0.2.1 *If* A *is diagonal-dominant, Gaussian elimination with* no *pivoting can be applied to solve* $A\mathbf{x} = \mathbf{b}$ *and a zero pivot will* not *be encountered. Thus, also,* A *is nonsingular.*

In other words, with diagonal-dominant systems, pivoting is not strictly necessary.

Proof Let the elements of A be denoted by α_{ij}. Now, α_{11} is certainly not zero, because the diagonal elements of a diagonal-dominant matrix are greater than something (the sum of the absolute values of the off-diagonal elements) that is nonnegative. Thus let us apply Gaussian elimination to zero all the elements below α_{11} in the first column (see Figure 0.2.1). If the new elements of A are denoted by β_{ij} (except the first row, which has not changed), then

$$\beta_{ij} = \alpha_{ij} - \frac{\alpha_{i1}\alpha_{1j}}{\alpha_{11}}. \tag{0.2.1}$$

We will show that the new matrix is still diagonal-dominant, and therefore $\beta_{22} \neq 0$.

We look at the ith row ($i > 1$) and bound the off-diagonal elements:

$$\sum_{\substack{j \neq 1 \\ j \neq i}} |\beta_{ij}| \leq \sum_{\substack{j \neq 1 \\ j \neq i}} |\alpha_{ij}| + \sum_{\substack{j \neq 1 \\ j \neq i}} |\alpha_{i1}/\alpha_{11}| \cdot |\alpha_{1j}|$$

$$< [|\alpha_{ii}| - |\alpha_{i1}|] + |\alpha_{i1}/\alpha_{11}|[|\alpha_{11}| - |\alpha_{1i}|]$$

$$= |\alpha_{ii}| - |\alpha_{i1}\alpha_{1i}/\alpha_{11}|$$

$$\leq |\alpha_{ii} - \alpha_{i1}\alpha_{1i}/\alpha_{11}| = |\beta_{ii}|.$$

$$
\begin{matrix}
\alpha_{11} & \cdots & \alpha_{1j} & \cdots \\[4pt]
0 & \boxed{\begin{matrix} \beta_{22} & & \cdot \\ & \cdot & \\ \cdots & & \beta_{ij} \\ & \cdot & \end{matrix}} \\
0 & \\
\cdot &
\end{matrix}
$$

Figure 0.2.1

This shows that the new matrix is still diagonal-dominant and thus $\beta_{22} \neq 0$. Applying the same argument again to the diagonal-dominant $N - 1$ by $N - 1$ "β" matrix (the box in Figure 0.2.1), we see that $\beta_{33} \neq 0$ after the subdiagonal elements in column 2 are zeroed, and so on. ∎

By changing the above "less than" inequalities to "less than or equal," we can show that "weak" diagonal dominance is also preserved during the Gaussian elimination without pivoting. Then a pivot element can be zero only if its entire row is zero, because it is greater than or equal to (in absolute value) the sum of the absolute values of the other elements in its row. Thus we have

Theorem 0.2.2 *If A is weakly diagonal-dominant, Gaussian elimination with no pivoting can be applied to solve $Ax = b$, and a zero pivot will only be encountered if A is singular.*

Now, if the original matrix is symmetric, $\alpha_{ij} = \alpha_{ji}$, then the $N - 1$ by $N - 1$ "β" matrix is still symmetric, for, by 0.2.1,

$$\beta_{ji} = \alpha_{ji} - \frac{\alpha_{j1}\alpha_{1i}}{\alpha_{11}} = \alpha_{ij} - \frac{\alpha_{1j}\alpha_{i1}}{\alpha_{11}} = \beta_{ij}.$$

If A is symmetric and weakly diagonal-dominant, then not only is β_{22} greater than or equal to (in absolute value) all other elements in its row, but also (by symmetry) greater than or equal to all other subdiagonal elements in its column. Thus

Theorem 0.2.3 *If A is weakly diagonal-dominant and symmetric, Gaussian elimination with no pivoting is identical to Gaussian elimination with partial pivoting.*

There is another class of matrices that arise frequently in this text that may not be quite diagonal-dominant but that still require no pivoting during elimination. These matrices are characterized by the following definition:

Definition A matrix A is called *positive-definite* if it is symmetric, and

$$\mathbf{x}^T A \mathbf{x} > 0$$

for all $\mathbf{x} \neq \mathbf{0}$.

Throughout this text, a vector is considered to be a column vector, i.e., an $N \times 1$ matrix. Thus its transpose is a row vector, so that in this definition $\mathbf{x}^T A \mathbf{x}$ means the dot product of $\mathbf{x}$ with $A\mathbf{x}$.

It is well known and easily proven (Problem 3a) that the eigenvalues of a symmetric matrix are all real. For a (symmetric) positive-definite matrix the above definition shows that they are also positive, for suppose $A\mathbf{x} = \lambda\mathbf{x}(\mathbf{x} \neq \mathbf{0})$. Then $0 < \mathbf{x}^T A \mathbf{x} = \lambda\mathbf{x}^T\mathbf{x} = \lambda\|\mathbf{x}\|_2^2$, so that $\lambda > 0$. It can also be shown that the converse is true, that is, if A is symmetric with positive eigenvalues, it is positive-definite (Problem 3b).

It will now be shown that pivoting is not necessary for positive-definite matrices.

Theorem 0.2.4 *If A is positive-definite, Gaussian elimination with* no *pivoting can be applied to solve $A\mathbf{x} = \mathbf{b}$, and a zero pivot will* not *be encountered. Thus, also, A is nonsingular.*

Proof First, if $\mathbf{e}_1 = (1, 0, \ldots, 0)$, then $0 < \mathbf{e}_1{}^T A \mathbf{e}_1 = \alpha_{11}$ so that certainly the first pivot is not zero. We want to show that after the subdiagonal elements in the first column are zeroed, the $N - 1$ by $N - 1$ "β" matrix (Figure 0.2.1) is still positive-definite, and $\beta_{22} > 0$, and so on.

Let us call $\mathbf{u} = (\alpha_{21}, \alpha_{31}, \ldots, \alpha_{N1})$. Then the original matrix can be drawn as shown below:

$$\begin{bmatrix} \alpha_{11} & \mathbf{u}^T \\ \mathbf{u} & A_1 \end{bmatrix},$$

where A_1 is the remaining $N - 1$ by $N - 1$ portion of A. After zeroing the subdiagonal elements in column one, the matrix has the form

$$\begin{bmatrix} \alpha_{11} & \mathbf{u}^T \\ \mathbf{0} & B_1 \end{bmatrix},$$

where the elements of B_1 are given by formula (0.2.1),

$$\beta_{ij} = \alpha_{ij} - \alpha_{i1}\alpha_{1j}/\alpha_{11} = \alpha_{ij} - u_{i-1}u_{j-1}/\alpha_{11},$$

so

$$B_1 = A_1 - \frac{\mathbf{u}\mathbf{u}^T}{\alpha_{11}}.$$

It was shown earlier that B_1 is symmetric. To show that it is also positive-definite, let **w** be any nonzero $N - 1$ vector. Then

$$\mathbf{w}^T B_1 \mathbf{w} = \mathbf{w}^T (A_1 - \mathbf{u}\mathbf{u}^T/\alpha_{11})\mathbf{w}$$

$$= [-\mathbf{u}^T\mathbf{w}/\alpha_{11} \quad \mathbf{w}^T]\begin{bmatrix} \alpha_{11} & \mathbf{u}^T \\ \mathbf{u} & A_1 \end{bmatrix}\begin{bmatrix} -\mathbf{u}^T\mathbf{w}/\alpha_{11} \\ \mathbf{w} \end{bmatrix}$$

$$= \mathbf{z}^T A \mathbf{z} > 0,$$

where **z** is the (nonzero) N vector whose first component is $-\mathbf{u}^T\mathbf{w}/\alpha_{11}$ and whose last $N - 1$ components are those of **w**. Since $\mathbf{w}^T B_1 \mathbf{w} > 0$ for arbitrary nonzero **w**, this proves that B_1 is positive-definite. Therefore $\beta_{22} > 0$ and, continuing the argument, it can be seen that a zero pivot cannot occur. ∎

Theorems 0.2.1 and 0.2.4 show that if A is either diagonal-dominant or positive-definite, Gaussian elimination without pivoting can be applied to it without fear of an *exactly* zero pivot, but we have already seen in Section 0.1 that a nearly zero pivot is almost as fatal as a zero pivot. Therefore, Problem 5 supplies bounds on how close to zero the pivots can be in these two cases.

For a full matrix A, it has been observed that pivoting involves a negligible amount of work, so why are we interested in avoiding it? The real reason is that for band and general sparse matrices (see Sections 0.4 and 0.5) pivoting not only complicates the programming but also results in a non-negligible decrease in efficiency.

0.3 The *LU* Decomposition

For applications involving time-dependent differential equations we frequently encounter the need to solve several linear systems (one or more each time step) that have the same matrix A but different right-hand sides. Our first inclination might be to find the inverse of A and multiply each right-hand side by A^{-1}. This is inefficient even when A is a full matrix, but it is especially unacceptable when A is a band matrix, because the inverse of a band matrix is generally full.

Instead, we will form what is called the "*LU* decomposition" of A; this will serve the same purpose as calculating an inverse but is much more efficient. In fact, when the first linear system is solved, the *LU* decomposition can be obtained with no extra cost.

Initially it will be assumed that no pivoting is done during the solution of $A\mathbf{x} = \mathbf{b}$ by Gaussian elimination. To obtain the LU decomposition, only one change from the normal algorithm is required: Whenever an element $\alpha_{ik}(i > k)$ is zeroed, instead of replacing α_{ik} by zero, the negative of the multiplier, $m_{ik} = -\alpha_{ik}/\alpha_{kk}$, is placed in this position, although it is understood that this position is now really zero. After it has been triangularized, the matrix A has the form shown below (for $N = 4$):

$$\begin{bmatrix} u_{11} & u_{12} & u_{13} & u_{14} \\ (-m_{21}) & u_{22} & u_{23} & u_{24} \\ (-m_{31}) & (-m_{32}) & u_{33} & u_{34} \\ (-m_{41}) & (-m_{42}) & (-m_{43}) & u_{44} \end{bmatrix}.$$

Then it will be shown that $A = LU$, where L and U are respectively the lower and upper triangular matrices given by

$$L = \begin{bmatrix} 1 & 0 & 0 & 0 \\ -m_{21} & 1 & 0 & 0 \\ -m_{31} & -m_{32} & 1 & 0 \\ -m_{41} & -m_{42} & -m_{43} & 1 \end{bmatrix}, \quad U = \begin{bmatrix} u_{11} & u_{12} & u_{13} & u_{14} \\ 0 & u_{22} & u_{23} & u_{24} \\ 0 & 0 & u_{33} & u_{34} \\ 0 & 0 & 0 & u_{44} \end{bmatrix}. \quad (0.3.1)$$

To see this, first verify that $L = M_1 M_2 M_3$ where

$$M_1 = \begin{bmatrix} 1 & 0 & 0 & 0 \\ -m_{21} & 1 & 0 & 0 \\ -m_{31} & 0 & 1 & 0 \\ -m_{41} & 0 & 0 & 1 \end{bmatrix}, \quad M_2 = \begin{bmatrix} 1 & 0 & 0 & 0 \\ 0 & 1 & 0 & 0 \\ 0 & -m_{32} & 1 & 0 \\ 0 & -m_{42} & 0 & 1 \end{bmatrix},$$

$$M_3 = \begin{bmatrix} 1 & 0 & 0 & 0 \\ 0 & 1 & 0 & 0 \\ 0 & 0 & 1 & 0 \\ 0 & 0 & -m_{43} & 1 \end{bmatrix},$$

so that $LU = M_1 M_2 M_3 U$. But notice that the effect of premultiplying U by M_3 is to subtract back the multiple (m_{43}) of row three, which was added to row four when element α_{43} was zeroed. In other words, $M_3 U$ is precisely what A looked like right before zeroing the subdiagonal element in column three. Next, premultiplying this by M_2 has the effect of subtracting back the multiples $(m_{32}$ and $m_{42})$ of row two, which were added to rows three and four in order to zero the subdiagonal elements in the second column. Finally, premultiplying by M_1 brings us all the way back to the original matrix A. Thus

$$LU = M_1 M_2 M_3 U = A.$$

Once the LU decomposition of A has been formed, to solve another system $A\mathbf{x} = \mathbf{b}$ only the following systems need to be solved:

$$L\mathbf{y} = \mathbf{b}$$

$$\text{then} \quad U\mathbf{x} = \mathbf{y}.$$

Both of these are triangular systems and hence require very little work ($O(N^2)$ operations) to solve, compared to the original problem $A\mathbf{x} = \mathbf{b}$.

A nice property of L and U is that if A is banded, so are L and U, as will be verified later, while A^{-1} may be full (see Problem 6).

If pivoting is allowed, this complicates things slightly, but the basic idea is to keep up with, via a permutation vector, the row switches done during the LU decomposition. Then when a new system $A\mathbf{x} = \mathbf{b}$ is solved, those same row switches must be done to $\mathbf{b}$ *before* $LU\mathbf{x} = \mathbf{b}$ is solved. (See Problem 2.)

For positive-definite matrices there is another factorization related to the LU factorization, called the "Cholesky" decomposition, which will be of interest later, when the eigenvalue problem is studied in Chapters 4 and 5.

Theorem 0.3.1 *If A is positive-definite, there exists a nonsingular lower triangular matrix L such that $A = LL^T$.*

Proof Since A is positive-definite, no pivoting is necessary during Gaussian elimination, so we can begin with the decomposition $A = L_1 U_1$ constructed as outlined above, where L_1 is a lower triangular matrix with 1s along the diagonal (see 0.3.1), and U_1 is an upper triangular matrix whose diagonal entries (the Gauss elimination pivots) were shown to be positive in the proof of Theorem 0.2.4. If the diagonal portion of U is called D, since these elements are positive, $D^{1/2}$ is a real matrix. Thus we can define $L = L_1 D^{1/2}$ and $U = D^{-1/2} U_1$, and clearly $A = LU$, where now both L and U have diagonal portion equal to $D^{1/2}$. Now, if it can be shown that $U = L^T$, the proof of Theorem 0.3.1 is complete, since L is lower triangular and, because its diagonal elements are positive, nonsingular.

By the symmetry of A, we have

$$LU = U^T L^T,$$

$$U^{-T} L = L^T U^{-1}.$$

Since the inverse of a lower triangular matrix is lower triangular, and the product of two lower triangular matrices is also lower triangular (and similarly for upper triangular matrices), the left-hand side is lower triangular and the right-hand side is upper triangular. Therefore each side must be equal to a diagonal matrix, which we will call E. Then $L^T U^{-1} = E$, so $EU = L^T$. Equating the diagonal portions of the two sides gives $ED^{1/2} = D^{1/2}$, so that E must be the identity matrix, and so $U = L^T$. ∎

0.4 Banded Linear Systems

Most of the linear systems encountered when solving differential equations are very large, if high accuracy is required, but fortunately they are also sparse, that is, most of their elements are zero. If the equations and unknowns are ordered properly, these matrices can generally be put into "banded" form, with all nonzero elements confined to a relatively narrow band around the main diagonal. If $\alpha_{ij} = 0$ when $|i - j| > L$, the matrix A is said to be banded, with "half bandwidth" equal to L.

When Gaussian elimination is applied to a band matrix (Figure 0.4.1), *without pivoting*, the zero elements outside the band remain zero, since multiples of row k are added to row i only when $|i - k| \leq L$. In addition, if the LU decomposition of A is formed, and the zeroed elements are replaced by the negatives of the multipliers, the elements outside the band are still zero after forward elimination and thus L and U are both banded, as previously mentioned.

A FORTRAN77 program to solve a banded linear system without pivoting is shown in Figure 0.4.3. The band matrix is stored in an $N \times 2L + 1$ rectangular array, whose columns are the diagonals of A. (Figure 0.4.2 illustrates this storage format for the case $N = 6$, $L = 2$.)

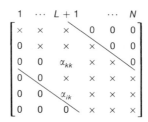

Figure 0.4.1

$$
\begin{array}{cc}
a_{11} & a_{12} \\
& a_{21}
\end{array}
\left[
\begin{array}{ccccccc}
a_{13} & a_{14} & a_{15} & 0 & 0 & 0 \\
a_{22} & a_{23} & a_{24} & a_{25} & 0 & 0 \\
a_{31} & a_{32} & a_{33} & a_{34} & a_{35} & 0 \\
0 & a_{41} & a_{42} & a_{43} & a_{44} & a_{45} \\
0 & 0 & a_{51} & a_{52} & a_{53} & a_{54} \\
0 & 0 & 0 & a_{61} & a_{62} & a_{63}
\end{array}
\right.
\begin{array}{cc}
 & \\
 & \\
 & \\
 & \\
a_{55} & \\
a_{64} & a_{65}
\end{array}
$$

Figure 0.4.2

There are triangles of elements in this array in the upper left-hand and lower right-hand corners that store undefined elements outside the matrix, but these are never referenced and thus cause no problems.

From Figure 0.4.3 it can be determined that if $1 \ll L$, the amount of computer time spent in all other parts of the code is negligible compared to that spent in DO loop 20, where a multiple of row K is added to row I. Most of the work is done, then, by

```
     DO 40 K=1,N-1
        DO 30 I=K+1,MIN(K+L,N)
           DO 20 J=K,MIN(K+L,N)
              A(I,JL) = A(I,JL) + AMUL*A(K,KL)
              .
20         CONTINUE
30      CONTINUE
40 CONTINUE
```

If it is also assumed that $L \ll N$, then the number of times the statement $A(I, JL) = A(I, JL) + AMUL*A(K, KL)$ is executed, that is, the number of multiplications done, is approximately (since "usually" $MIN(K + L, N) = K + L$)

$$
\sum_{K=1}^{N-1} \sum_{I=K+1}^{K+L} \sum_{J=K}^{K+L} 1 = (N-1)L(L+1) \approx NL^2.
$$

On the other hand, if pivoting is allowed, some of the elements outside the original band may become nonzero during the elimination. In Figure 0.4.2, for example, if the first and third rows are interchanged to bring the element $a_{31} = \alpha_{31}$ to the pivot position ($a_{13} = \alpha_{11}$) in the first column, then the new first row may have nonzeros extending two positions past the limit of the original band. This is the worst that can happen, however, and in general, if pivoting is allowed, there are L additional diagonals added on to the top of the original band and none added on below the band. This increases the

```
      SUBROUTINE LBAND(A,X,B,IA,N,L)
      IMPLICIT DOUBLE PRECISION(A-H,O-Z)
C
C     ARGUMENT DESCRIPTIONS
C
C     A -(INPUT) A IS AN N BY 2*L+1 ARRAY CONTAINING THE BAND MATRIX IN
C         BAND STORAGE MODE, ACTUALLY DIMENSIONED IA BY 2*L+1.
C         (OUTPUT) A IS DESTROYED.
C     X -(OUTPUT) X IS THE SOLUTION VECTOR OF LENGTH N.
C     B -(INPUT) B IS THE RIGHT HAND SIDE VECTOR OF LENGTH N.
C     IA-(INPUT) IA IS THE FIRST DIMENSION OF MATRIX A, AS ACTUALLY
C         DIMENSIONED IN THE CALLING PROGRAM (IA.GE.N).
C     N -(INPUT) N IS THE NUMBER OF EQUATIONS AND NUMBER OF UNKNOWNS
C         IN THE LINEAR SYSTEM.
C     L -(INPUT) L IS THE HALF-BANDWIDTH, DEFINED AS THE MAXIMUM VALUE
C         OF ABS(I-J) SUCH THAT AIJ IS NONZERO.
C
      DIMENSION A(IA,*),X(N),B(N)
C                               COPY B ONTO X
      DO 10 I=1,N
   10 X(I) = B(I)
C                               BEGIN FORWARD ELIMINATION
      DO 40 K=1,N-1
      IF (A(K,L+1).EQ.0.D0) GO TO 70
      IL = L
      DO 30 I=K+1,MIN(K+L,N)
         AMUL = -A(I,IL)/A(K,L+1)
         JL = IL
         KL = L+1
         IL = IL-1
         IF (AMUL.EQ.0.D0) GO TO 30
C                               ADD AMUL TIMES ROW K TO ROW I
         DO 20 J=K,MIN(K+L,N)
            A(I,JL) = A(I,JL)+AMUL*A(K,KL)
            JL = JL+1
            KL = KL+1
   20    CONTINUE
         X(I) = X(I)+AMUL*X(K)
   30 CONTINUE
   40 CONTINUE
      IF (A(N,L+1).EQ.0.D0) GO TO 70
C                               BEGIN BACK SUBSTITUTION
      X(N) = X(N)/A(N,L+1)
      DO 60 K=N-1,1,-1
      KL = L+1
      DO 50 J=K+1,MIN(K+L,N)
         KL = KL+1
         X(K) = X(K)-A(K,KL)*X(J)
   50 CONTINUE
      X(K) = X(K)/A(K,L+1)
```

```
      60 CONTINUE
         RETURN
C                                     MATRIX IS SINGULAR, OR PIVOTING MUST
C                                     BE DONE.
      70 PRINT 80
      80 FORMAT (23H ZERO PIVOT ENCOUNTERED)
         RETURN
         END
```

Figure 0.4.3
Banded System Solver (No Pivoting)

```
      SUBROUTINE TRI(A,B,C,X,F,N)
      IMPLICIT DOUBLE PRECISION(A-H,O-Z)
C
C     ARGUMENT DESCRIPTIONS
C
C     A,B,C-(INPUT) A,B,C ARE VECTORS OF LENGTH N HOLDING THE
C           SUBDIAGONAL, DIAGONAL, AND SUPERDIAGONAL OF THE TRIDIAGONAL
C           MATRIX IN THE LINEAR SYSTEM.
C           (OUTPUT) A,B,C ARE DESTROYED.
C         X-(OUTPUT) X IS THE SOLUTION VECTOR OF LENGTH N.
C         F-(INPUT) F IS THE RIGHT HAND SIDE VECTOR OF LENGTH N.
C         N-(INPUT) N IS THE NUMBER OF EQUATIONS AND NUMBER OF UNKNOWNS.
C
      DIMENSION A(N),B(N),C(N),X(N),F(N)
C                                     COPY F ONTO X
      DO 5 I=1,N
    5 X(I) = F(I)
C                                     BEGIN FORWARD ELIMINATION
      DO 10 K=1,N-1
         IF (B(K).EQ.0.DO) GO TO 20
         AMUL = -A(K+1)/B(K)
         B(K+1) = B(K+1)+AMUL*C(K)
         X(K+1) = X(K+1)+AMUL*X(K)
   10 CONTINUE
C                                     BACK SUBSTITUTION
      IF (B(N).EQ.0.DO) GO TO 20
      X(N) = X(N)/B(N)
      DO 15 K=N-1,1,-1
         X(K) = (X(K)-C(K)*X(K+1))/B(K)
   15 CONTINUE
      RETURN
C                                     MATRIX IS SINGULAR, OR PIVOTING
C                                     MUST BE DONE.
   20 PRINT 25
   25 FORMAT (23H ZERO PIVOT ENCOUNTERED)
      RETURN
      END
```

Figure 0.4.4
Tridiagonal System Solver (No Pivoting)

$$
\begin{bmatrix}
b_1 & c_1 & d_1 \\
a_2 & b_2 & c_2 & d_2 \\
& & \cdot & \cdot & \cdot \\
& & a_k & b_k & c_k & d_k \\
& & & a_{k+1} & b_{k+1} & c_{k+1} & d_{k+1} \\
& & & & \cdot & \cdot & \cdot \\
& & & & & a_N & b_N
\end{bmatrix}
\begin{bmatrix}
x_1 \\ x_2 \\ \cdot \\ x_k \\ x_{k+1} \\ \cdot \\ x_N
\end{bmatrix}
=
\begin{bmatrix}
f_1 \\ f_2 \\ \cdot \\ f_k \\ f_{k+1} \\ \cdot \\ f_N
\end{bmatrix}
$$

Figure 0.4.5

amount of storage required by approximately 50%. The total work is doubled, approximately, since the DO loop 20 will now have limits of K to $MIN(K + 2*L, N)$. In any case, the storage and operation count are still much less than for a full matrix. If L is constant, both are proportional to N.

In one-dimensional problems, tridiagonal systems (banded systems with $L = 1$) are encountered frequently. Because of the importance of these systems, a FORTRAN77 subroutine that solves tridiagonal linear systems without pivoting is exhibited in Figure 0.4.4. The three diagonals of the tridiagonal matrix are assumed to be stored in three arrays **a**, **b**, **c** (see Figure 0.4.5).

If pivoting is done, there is an extra work array **d** representing the diagonal immediately above **c** (see Figure 0.4.5). The fourth array **d** is initially zero. When the forward elimination reaches column k, $|b_k|$ is compared with $|a_{k+1}|$, the only other potential pivot. If $|a_{k+1}|$ is larger, rows k and $k + 1$ are interchanged to move the larger element into the pivot position. This will generally introduce a nonzero element into position d_k, and thus the fourth diagonal may become nonzero during the elimination. Figure 0.4.6 shows a

```
        SUBROUTINE TRI(A,B,C,X,F,N)
        IMPLICIT DOUBLE PRECISION(A-H,O-Z)
C
C       ARGUMENT DESCRIPTIONS
C
C       A,B,C-(INPUT), A,B,C ARE VECTORS OF LENGTH N HOLDING THE
C            SUBDIAGONAL, DIAGONAL, AND SUPERDIAGONAL OF THE TRIDIAGONAL
C            MATRIX IN THE LINEAR SYSTEM.
C            (OUTPUT) A,B,C ARE DESTROYED.
C       X-(OUTPUT) X IS THE SOLUTION VECTOR OF LENGTH N.
C       F-(INPUT) F IS THE RIGHT HAND SIDE VECTOR OF LENGTH N.
C       N-(INPUT) N IS THE NUMBER OF EQUATIONS AND NUMBER OF UNKNOWNS.
C
        PARAMETER (NMAX=1000)
        DIMENSION A(N),B(N),C(N),D(NMAX),X(N),F(N)
        IF (N.GT.NMAX) GO TO 30
```

```
C                              COPY F ONTO X
      DO 5 I=1,N
         X(I) = F(I)
    5    D(I) = 0.0
C                              BEGIN FORWARD ELIMINATION
      DO 10 K=1,N-1
         IF (ABS(A(K+1)).GT.ABS(B(K))) THEN
C                              SWITCH ROWS K AND K+1
            TEMP = B(K)
            B(K) = A(K+1)
            A(K+1) = TEMP
            TEMP = C(K)
            C(K) = B(K+1)
            B(K+1) = TEMP
            TEMP = X(K)
            X(K) = X(K+1)
            X(K+1) = TEMP
            IF (K.LT.N-1) THEN
               TEMP = D(K)
               D(K) = C(K+1)
               C(K+1) = TEMP
            ENDIF
         ENDIF
         IF (B(K).EQ.0.D0) GO TO 20
         AMUL = -A(K+1)/B(K)
         B(K+1) = B(K+1)+AMUL*C(K)
         X(K+1) = X(K+1)+AMUL*X(K)
         IF (K.LT.N-1) THEN
            C(K+1) = C(K+1)+AMUL*D(K)
         ENDIF
   10 CONTINUE
C                              BACK SUBSTITUTION
      IF (B(N).EQ.0.D0) GO TO 20
      X(N) = X(N)/B(N)
      X(N-1) = (X(N-1)-C(N-1)*X(N))/B(N-1)
      DO 15 K=N-2,1,-1
         X(K) = (X(K)-C(K)*X(K+1)-D(K)*X(K+2))/B(K)
   15 CONTINUE
      RETURN
C                              THE MATRIX IS SINGULAR
   20 PRINT 25
   25 FORMAT (23H THE MATRIX IS SINGULAR)
      RETURN
C                              WORKSPACE OVERFLOW (ARRAY D)
   30 PRINT 35
   35 FORMAT (42H INCREASE PARAMETER NMAX IN SUBROUTINE TRI)
      RETURN
      END
```

Figure 0.4.6
Tridiagonal System Solver (Partial Pivoting)

FORTRAN77 program that solves a tridiagonal system with partial pivoting.

0.5 Sparse Direct Methods

If a linear system has a sparse matrix, it ought to be possible in the Gaussian
elimination process to take advantage of the fact that most of the matrix
elements are zero. If the unknowns and equations are ordered randomly,
however, most of the zero elements may "fill-in," that is, become nonzero
(temporarily, in the case of subdiagonal elements) during the elimination,
with the result that the storage and execution time required are scarcely
better than if the matrix were full.

Ordering the unknowns and equations so that a sparse matrix has a small
bandwidth is one way to take advantage of sparseness during Gaussian
elimination; it is surely the easiest technique to understand and program, but
it is not necessarily the most efficient. There are other orderings that, while
not producing a small bandwidth, may result in less fill-in (which means less
storage and generally a lower operation count) than a band ordering. The
solution of a linear system using these "sparse matrix" orderings is typically
very complicated to implement efficiently, but a brief introduction to the
ideas involved will be presented here.

For all the linear systems generated by the finite difference and (Galerkin)
finite element discretization methods that will be studied in this text, there is a
natural association between the unknowns and the equations. That is, to each
unknown there is an equation associated with it, and when the unknowns and
their associated equations are assigned the same ordering, the coefficient
matrix always has a symmetric nonzero structure, even though the matrix
itself may be nonsymmetric. Further, the diagonal entries will (almost)
always be nonzero, and the matrix is often positive definite, diagonal
dominant, or otherwise composed so that pivoting is not essential. Thus we
will always use the natural ordering of the equations once an ordering for the
unknowns has been chosen. We will assume that no pivoting is done during
the elimination, which ensures that symmetry in the nonzero structure is
preserved during elimination, in the uneliminated portion of the matrix (the
"β" portion of the matrix in Figure 0.2.1). If these well-justified assumptions
are not made, the sparse methods to be discussed here become much more
complex and (usually) inefficient.

One of the basic ideas motivating the orderings of the unknowns employed by efficient sparse matrix methods is the principle that elimination of unknowns corresponding to rows with many nonzeros, counting those generated by earlier fill-in, should be delayed as long as possible. Figures 0.5.1 and 0.5.2 illustrate this principle. Elements that are nonzero in the original matrix are marked by (X) and elements that are originally zero but become nonzero due to fill-in are marked by (—). In Figure 0.5.1, the nonzero elements on the top row cause the entire matrix to fill-in (no pivoting is assumed), while when the first unknown and equation are moved to the last position (Figure 0.5.2) there is no fill-in at all. In general, rows with many nonzeros near the bottom of the matrix do not have "time" to generate much further fill-in.

```
X  X  X  X  X  X  X  X  X  X            X                          X
X  X  —  —  —  —  —  —  —  —             X                          X
X  —  X  —  —  —  —  —  —  —                X                       X
X  —  —  X  —  —  —  —  —  —                   X                    X
X  —  —  —  X  —  —  —  —  —                      X                 X
X  —  —  —  —  X  —  —  —  —                         X              X
X  —  —  —  —  —  X  —  —  —                            X           X
X  —  —  —  —  —  —  X  —  —                               X        X
X  —  —  —  —  —  —  —  X  —                                  X  X  X
X  —  —  —  —  —  —  —  —  X             X  X  X  X  X  X  X  X  X  X
```

Figure 0.5.1   **Figure 0.5.2**

If the "degree" of unknown number k is defined to be the number of nonzeros to the right of the diagonal in row k, or equivalently (since the matrix is assumed to have symmetric nonzero structure), the number of nonzeros below the diagonal in column k, then the quantity of fill-in generated by the elimination of unknown k can be seen to be at most the square of the degree of that unknown (cf. Figure 0.5.1). In the *minimal degree* algorithm, the unknowns are ordered in such a way that after the first $k - 1$ unknowns have been eliminated (the first $k - 1$ columns have been zeroed), the next unknown (k) chosen to be eliminated is the one whose current degree is minimal, compared to the remaining unknowns $k + 1, k + 2, \ldots$. Thus, of all the remaining unknowns, the one chosen will generate the least amount of fill-in *at that stage* when it is eliminated.

It is not necessarily true that deciding which unknown to eliminate next, based on which will generate the least fill-in at that stage, will produce the least possible fill-in in the end, but this ordering seems reasonable, and it is in keeping with the principle that elimination of unknowns corresponding to rows with many nonzeros should be delayed as long as possible. And, unfortunately, finding the ordering that will produce the least fill-in in the end would require an absolutely prohibitive amount of work.

It is not obvious that the same cannot be said of the minimal degree ordering itself, that is, that it is prohibitively costly to determine. However, algorithms do exist that are able to find the minimal degree ordering in a reasonable amount of time. Even after this ordering has been found, doing the Gaussian elimination in such a way that only elements that are nonzero, or will become nonzero due to fill-in, are stored and operated upon is obviously much more complicated than in the case of a band ordering.

For most finite difference and finite element generated matrices, it is possible to determine an approximately minimal degree ordering based on the geometry of the problem. The *nested dissection* algorithms are the best known algorithms of this type. A complete description of these orderings is beyond the scope of this text, but the following example provides some insights.

Many two-dimensional finite difference problems lead to matrices whose nonzero structure can be represented by a graph such as that in Figure 0.5.3. In this graph each node represents an unknown, and elements a_{ij}, a_{ji} ($i \neq j$) of the matrix are nonzero if and only if nodes i and j are connected by a line. The unknowns in Figure 0.5.3 are ordered in such a way as to produce a small

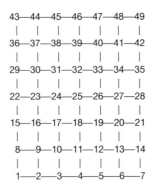

Figure 0.5.3
Band ordering

bandwidth. Figure 0.5.5 shows the nonzero elements and fill-in that result using this ordering.

An ordering that displays the principle behind the nested dissection algorithms is shown in Figure 0.5.4. Here the graph is divided, by two lines of nodes in **boldface**, into four subregions. The unknowns that subdivide the region (those in boldface) are ordered last, after all others. As can be seen by studying Figure 0.5.6, this results in postponing much of the fill-in until the last few rows, corresponding to the boldface nodes. Intuitively, this is because after the unknowns 1–9 in the first subregion have been eliminated, those in the other subregions still have the same degrees as they had originally, while the degrees of some of the divider unknowns have increased. Thus the unknowns to be eliminated next are chosen from another subregion.

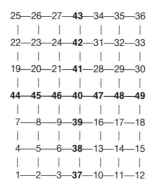

Figure 0.5.4
Dissection Ordering

The total fill-in (including elements originally nonzero) is reduced from 649 for the band ordering (Figure 0.5.5) to 579 for this ordering (Figure 0.5.6), and the operation count (assuming full advantage is taken of zero elements) is reduced from 1994 to 1619 for the forward elimination.

A double dissection ordering would further subdivide each of the four subregions, ordering the unknowns on these secondary divider lines (in the lower left-hand subregion, for example, the secondary divider consists of unknowns 2, 4, 5, 6, and 8) next-to-last, before only the unknowns on the primary dividers (in boldface above). The total fill-in is reduced further to 519 and the operation count is reduced to 1337. If the number of nodes were larger, a nested dissection algorithm would further subdivide the region.

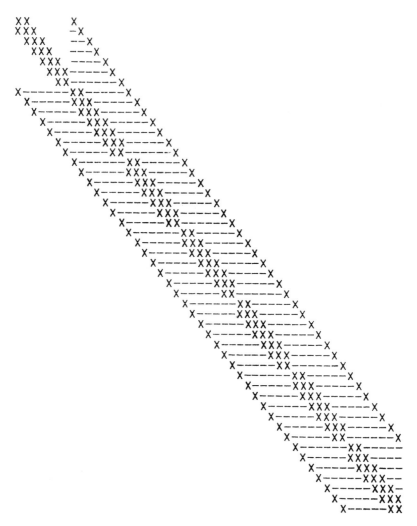

Figure 0.5.5
Fill-in Using a Band Ordering
X = nonzero in original matrix   − = nonzero due to fill-in

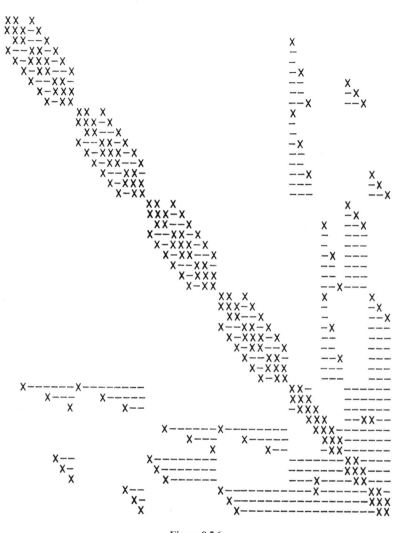

Figure 0.5.6
Fill-in Using Dissection Ordering
X = nonzero in original matrix   − = nonzero due to fill-in

Although for this problem these orderings yield only a minimal gain over the band ordering, for $N \times N$ grids of the above form, with N large, it has been shown that a nested dissection ordering requires only $O(N^2 \log N)$ storage and $O(N^3)$ work, as compared to $O(N^3)$ and $O(N^4)$, respectively, for the band solver. However, there is so much overhead involved in all these sparse direct methods that a fairly large problem is required before they are actually faster than band solvers.

For more detail on sparse Gaussian elimination, see Duff, Erisman, and Reid (1987).

0.6 Problems

1. In Section 0.2 it was shown that if Gaussian elimination is performed on a symmetric matrix without pivoting, then, after the subdiagonal portions of the first k columns have been zeroed, the remaining $N - k \times N - k$ lower right corner of the matrix is still symmetric. Modify the subroutine LINEQ of Figure 0.1.2 so that no pivoting is allowed, and so that elements below the diagonal are never referenced or modified, it being understood that $A(I, J)$ for $J < I$ is equal to $A(J, I)$ until it has been zeroed. The resulting subroutine will then execute approximately twice as fast as the old version when A is symmetric and requires no pivoting (e.g., if positive definite).

Test the new subroutine on the positive definite system

$$\begin{bmatrix} 7 & -1 & -2 & -3 \\ -1 & 7 & -3 & -2 \\ -2 & -3 & 7 & -1 \\ -3 & -2 & -1 & 7 \end{bmatrix} \begin{bmatrix} x_1 \\ x_2 \\ x_3 \\ x_4 \end{bmatrix} = \begin{bmatrix} 1 \\ 1 \\ 1 \\ 1 \end{bmatrix},$$

which has solution $(1, 1, 1, 1)$. To make sure you never reference the lower triangle, initialize the subdiagonal elements of A to zero before calling the linear equation solver.

It is clearly possible to cut the amount of stortage in half by storing only the diagonal and upper triangle, but this requires a more complicated and inconvenient storage format for A than the $N \times N$ array used by LINEQ.

2. Write two FORTRAN subroutines:

   a. SUBROUTINE DECOMP(A,N,IPERM)

   DECOMP should reduce the $N \times N$ matrix A to upper triangular form using Gaussian elimination with partial pivoting (use LINEQ in Figure 0.1.2 as a starting point), replacing the zeroed element $A(I, J)$ ($J < I$) not by zero, but by the negative of the multiplier used to zero it. Thus on output, the diagonal and upper triangle of A contain the matrix U, and the lower triangle of A contains the lower triangle of L (recall that the diagonal elements of L are 1s), where LU is equal to the matrix that results when the rows of A are permuted as dictated by the partial pivoting. IPERM should be an integer vector that keeps up with the pivoting done during the elimination process. IPERM(K) should be initialized to K (for $K = 1, \ldots, N$), and, each time rows I and J of A are switched, the values of IPERM(I) and IPERM(J) should also be interchanged. (Note: In Figure 0.1.2, when two rows are interchanged, only the nonzero portions are swapped, but now the entire rows must be swapped.)

   b. SUBROUTINE SOLVE(A,N,IPERM,X,B)

   SOLVE solves $Ax = \mathbf{b}$, after DECOMP has been called to decompose A. On input, A and IPERM are as output by DECOMP, and B is the right-hand vector $\mathbf{b}$. On output, X is the solution vector $\mathbf{x}$, calculated by solving $LUx = \mathbf{c}$, where $\mathbf{c}$ is the vector that results when the elements of $\mathbf{b}$ are reordered in the same manner as the rows of A were reordered by subroutine DECOMP. That is, $c_i = B(\text{IPERM}(i))$. $LUx = \mathbf{c}$ is solved by solving first the lower triangular system $Ly = \mathbf{c}$ using forward substitution, then the upper triangular system $Ux = \mathbf{y}$ using back substitution. SOLVE must not alter the elements of A or IPERM.

To test these routines, call DECOMP once to decompose A, where

$$A = \begin{bmatrix} 1 & 2 & 3 \\ 3 & 6 & 8 \\ -2 & -5 & 4 \end{bmatrix},$$

and then call SOLVE twice, to solve $Ax = \mathbf{b}_1$ and $Ax = \mathbf{b}_2$, where

$$\mathbf{b}_1 = (5, 14, 0), \qquad \mathbf{b}_2 = (6, 17, -3)$$

3. a. Show that all eigenvalues of a symmetric matrix are real.

   Hint: If $Ax = \lambda x$, show that $\lambda x^T \bar{x} = \bar{\lambda} x^T \bar{x}$, where $\bar{x}$ is the complex conjugate of the eigenvector x.

   b. Show that if A is symmetric, and thus has real eigenvalues, then

$$\lambda_{min} = \min_{\mathbf{x} \neq \mathbf{0}} \frac{\mathbf{x}^T A \mathbf{x}}{\mathbf{x}^T \mathbf{x}},$$

   where λ_{min} is the (algebraically) smallest eigenvalue of A.
   You may use a fundamental theorem of linear algebra that states that for symmetric A, there exists a unitary matrix P (i.e., $P^{-1} = P^T$) such that $P^{-1}AP = D$, where D is a diagonal matrix containing the eigenvalues of A.

   It follows immediately from this result that if A is symmetric and has all positive eigenvalues, A is positive-definite by our definition.

4. Show that if A is symmetric and diagonal-dominant and has positive diagonal entries, it is positive-definite.
   Hint: $A - \lambda I$ is diagonal dominant for $\lambda \leq 0$.
   On the other hand, show by a counterexample that a positive-definite matrix need not be diagonal-dominant.

5. a. Show that if A is diagonal-dominant with "margin" $s > 0$, that is,

$$|\alpha_{ii}| \geq s + \sum_{j \neq i} |\alpha_{ij}|, \qquad 1 \leq i \leq N,$$

   then, if Gaussian elimination is performed on A without pivoting, all resulting pivots will be greater than or equal to s in absolute value.
   Hint: Modify the proof of Theorem 0.2.1 to show that after the subdiagonal elements in the first column have been zeroed, the remaining "β" matrix is diagonal-dominant with "margin" s, and therefore $|\beta_{22}| \geq s$.
   b. Show that if A is positive-definite, with smallest eigenvalue λ_{min}, then if Gaussian elimination is performed on A without pivoting, all resulting pivots will be greater than or equal to λ_{min}.
   Hint: Modify the proof of Theorem 0.2.4, using the result of Problem 3b, to show that

$$\mathbf{w}^T B_1 \mathbf{w} \geq \lambda_{min} \mathbf{w}^T \mathbf{w}$$

   for any $\mathbf{w}$, and therefore the smallest eigenvalue of B_1 is greater than or equal to λ_{min}. Then show that the diagonal elements of a positive-definite matrix are greater than or equal to its smallest eigenvalue.

6. Find the LU decomposition and the inverse of the tridiagonal matrix

$$
\begin{bmatrix}
1 & 1 & 0 & 0 & 0 & 0 \\
1 & 2 & 1 & 0 & 0 & 0 \\
0 & 1 & 2 & 1 & 0 & 0 \\
0 & 0 & 1 & 2 & 1 & 0 \\
0 & 0 & 0 & 1 & 2 & 1 \\
0 & 0 & 0 & 0 & 1 & 2
\end{bmatrix}.
$$

Notice that the inverse is full, while L and U $(=L^T)$ are bidiagonal.

1
Initial Value Ordinary Differential Equations

1.0 Introduction

Differential equations are often divided into two classes, ordinary and partial, according to the number of independent variables, and studied separately. A more meaningful division, however, is between initial value problems, which usually model time-dependent phenomena, and boundary value problems, which generally model steady-state systems. The differences between initial value and boundary value problems, and the similarities within each of these classes, are even more striking when numerical methods for these problems are considered.

The identification of initial value problems with time-dependency and of boundary value problems with a steady-state condition is helpful in under-standing some of the differences in the properties of these two types of problems. For example, the solution at any time of a time-dependent problem logically depends only on what has gone on before and not on future events. For a steady-state problem, on the other hand, the solution values at different spatial points may be interdependent. Thus it is not surprising that a time-dependent (initial value) problem can be solved numerically by marching forward in time from the given initial values, while a system of simultaneous algebraic equations must generally be solved to find the solution to a steady-state (boundary value) problem.

It is also clear why initial value problems almost always have unique solutions, while boundary value problems sometimes have many or no solutions. Consider, for example, the general second order equation $m u_{tt} = \mathbf{f}(t, \mathbf{u}, \mathbf{u}_t)$, which may be thought of as modeling Newton's second law, applied to an object whose coordinates are given by the vector $\mathbf{u}(t)$. With initial conditions $\mathbf{u}(0) = \mathbf{u_0}$, $\mathbf{u}_t(0) = \mathbf{u_1}$, under very reasonable smoothness assumptions on the force field $\mathbf{f}$, this problem will always have a unique solution. To find it, one only needs to create the force field described by $\mathbf{f}$ and release the object with the prescribed initial position and velocity. The object will find the solution even if we cannot! With boundary conditions $\mathbf{u}(0) = \mathbf{u_0}$, $\mathbf{u}(1) = \mathbf{u_1}$, there is no guarantee that a unique solution exists, as we are now requiring that the object travel from point $\mathbf{u_0}$ to point $\mathbf{u_1}$ in a prescribed time, under the specified force field, a requirement that may be impossible to satisfy or that may be satisfied by many trajectories. Note that we have just given an example of a time-dependent boundary value problem!

The problem to be studied in this chapter is the first-order initial value ordinary differential equation problem

$$\frac{du}{dt} = f(t, u), \qquad u(0) = u_0.$$

All of the methods and results of this chapter generalize to systems of first-order equations ($\mathbf{u}' = \mathbf{f}(t, \mathbf{u})$, $\mathbf{u}(0) = \mathbf{u_0}$) in a straightforward manner, by simply allowing u, f, u_0 and their approximations to be vector quantities. The introduction of additional dependent variables does not have any of the complicating effects that the introduction of additional independent variables has.

Furthermore, higher-order initial value problems are easily converted into first-order systems. For example, the second-order problem

$$\frac{d^2 u}{dt^2} = g\left(t, u, \frac{du}{dt}\right), \qquad u(0) = A, \qquad \frac{du}{dt}(0) = B$$

can be reduced to a system of two first-order equations by introducing the auxiliary variable $v \equiv du/dt$:

$$\frac{du}{dt} = v, \qquad\qquad u(0) = A,$$

$$\frac{dv}{dt} = g(t, u, v), \qquad v(0) = B.$$

Thus it is sufficient to study only the single first-order equation $u' = f(t, u)$.

1.1 Euler's Method

Since by definition $(du/dt)(t) = \lim_{h\to 0}(u(t + h) - u(t))/h$, the simplest and most obvious approach to solving $du/dt = f(t, u)$ is to approximate it by

$$\frac{u(t + h) - u(t)}{h} \approx f(t, u(t)), \qquad (1.1.1)$$

where h is a small but nonzero stepsize. (The calculus student learns to take limits to calculate derivatives and definite integrals, while in numerical analysis the student learns to stop short of the limit!)

Starting with the given initial value $u(0) = u_0$, we can use 1.1.1 to approximate u at $t = h$, then $t = 2h$, and so on. Calling $t_k = kh$ and letting $U(t_k)$ represent the approximation to $u(t_k)$ (U is only defined at the points t_k), we have the "Euler" approximation

$$U(t_{k+1}) = U(t_k) + hf(t_k, U(t_k)),$$
$$U(t_0) = u_0.$$

As a first example, consider the initial value problem

$$\frac{du}{dt} = 2t(1 + u^2), \qquad u(0) = 0, \qquad (1.1.2)$$

which has the true solution $u(t) = \tan(t^2)$. A FORTRAN segment that solves this problem using Euler's method with $h = 0.02$ is as follows:

```
      H = 0.02
      U = 0
      T = 0
      DO 10 K=1,25
         U = U + H*2*T*(1+U**2)
         T = T + H
         PRINT T,U
   10 CONTINUE
```

This program produced the value $U(0.5) = 0.244243$, which gives an error of 0.011099 at this point. Decreasing h by a factor of two, to 0.01, reduced the error to 0.005573 at the same point, almost exactly a twofold reduction. Thus the error appears to be proportional to h.

As a second example, consider the problem

$$\frac{du}{dt} = \alpha u, \qquad u(0) = P,$$

which has the true solution $u(t) = Pe^{\alpha t}$. Here $u(t)$ can be considered to be the balance of a savings account with initial balance P and a (continuously compounded) annual interest rate of α. The Euler method approximation satisfies

$$U(t_{k+1}) = U(t_k) + h\alpha U(t_k), \qquad U(0) = P.$$

$U(t_{k+1})$ can be found explicitly from

$$U(t_{k+1}) = (1 + h\alpha)U(t_k) = (1 + h\alpha)^2 U(t_{k-1}) = \cdots = (1 + h\alpha)^{k+1}U(t_0)$$

or

$$U(t_k) = (1 + h\alpha)^k P,$$

which is precisely the account balance resulting from compounding the interest at intervals of h years, rather than continuously.

For this problem, then, the Euler error can be directly displayed at a fixed point, say $t = t_k$ ($k = t/h$):

$$U(t) - u(t) = P(1 + h\alpha)^{t/h} - Pe^{\alpha t}$$

$$= P[\{(1 + h\alpha)^{1/(h\alpha)}\}^{\alpha t} - e^{\alpha t}].$$

As $h \to 0$, the term in the curved brackets converges to e, and the error goes to zero. So, for this example, what was only suspected in the first example has been proven—that the Euler approximation converges to the true solution as $h \to 0$.

1.2   Truncation Error, Stability, and Convergence

By expanding $u(t + h)$ in a Taylor series (with remainder) about t, it can be seen that

$$\frac{u(t + h) - u(t)}{h} = \frac{u(t) + hu'(t) + \frac{1}{2}h^2 u''(\xi) - u(t)}{h} = u'(t) + \frac{1}{2}hu''(\xi)$$

$$(t \le \xi \le t + h),$$

so that the left-hand side of the Euler equation 1.1.1 is an $O(h)$ approximation to the derivative it replaced, provided the second derivative of the solution u

is bounded in the time interval of interest. As h goes to zero, the Euler equation better and better represents the original differential equation; that is, it is "consistent" with the differential equation. A measure of the consistency of a finite difference approximation with a differential equation is given by the "truncation error," defined as follows:

Definition   The *truncation error* is the amount by which the solution of the differential equation fails to satisfy the approximate (finite difference) equation, in "normalized" form. ("Normalized" means in a form such that the terms in the difference equation approximate the corresponding terms in the differential equation.)

Note that $[U(t + h) - U(t)]/h = f(t, U(t))$ is Euler's equation in normalized form. One could multiply this equation by h (or h^{100}) and obtain an equivalent equation with a quite different truncation error, if no normalization were required.

Now consistency can be defined more formally.

Definition   An approximate (finite difference) method is *consistent* with the differential equation if the truncation error goes to zero as the stepsize h goes to zero.

The truncation error is a tool that will be used heavily throughout the study of finite difference approximations, because it is a measure of consistency that is easily calculated. For the Euler method, for example, the truncation error is obtained by plugging the exact solution of $u' = f(t, u)$ into the (normalized) Euler equation:

$$T \equiv \frac{u(t + h) - u(t)}{h} - f(t, u(t)) = \frac{u(t + h) - u(t)}{h} - u'(t)$$

$$= u'(t) + \tfrac{1}{2}hu''(\xi) - u'(t) = \tfrac{1}{2}hu''(\xi) \qquad (t \leq \xi \leq t + h).$$

The truncation error is $O(h)$ and, therefore, the Euler method is a consistent approximation.

Consistency does not automatically guarantee convergence, however. Consider, for example, the following method for approximating $u' = f(t, u)$:

$$\frac{2u(t + h) + 3u(t) - 6u(t - h) + u(t - 2h)}{6h} \approx f(t, u(t)). \qquad (1.2.1)$$

By expanding $u(t + h)$, $u(t - h)$, and $u(t - 2h)$ in Taylor series about t, it can be verified that the left-hand side of 1.2.1 is equal to $u'(t) + h^3 u^{iv}(\xi)/12$, where $t - 2h \leq \xi \leq t + h$, so the truncation error for this method is $O(h^3)$. Not only is this method consistent with the differential equation $u' = f(t, u)$, but it is a higher order approximation to it than is the Euler equation. Therefore it might be expected that

$$U(t + h) = -1.5U(t) + 3U(t - h) - 0.5U(t - 2h) + 3hf(t, U(t)), \qquad (1.2.2)$$

which is obtained by solving 1.2.1 for $u(t + h)$ and calling the approximate solution U, would yield a more accurate solution than the Euler method. So 1.2.2 was applied to the problem $u' = 2t(1 + u^2)$, $u(0) = 0$ (cf. 1.1.2), previously solved using Euler's method, with stepsizes of $h = 0.02$ and $h = 0.01$.

There is a slight problem here, in that 1.2.2 can only be used once $U(0)$, $U(h)$, and $U(2h)$ are known. Obviously, $U(0) = 0$, but values for the next two starting points are not easily available. They could be obtained using another method such as one of the Runge–Kutta methods (Section 1.6), but for now we will simply use exact starting values, since the exact solution is known $(u(t) = \tan(t^2))$, so that the bad results that will be obtained will not be blamed on the starting values.

The results in Table 1.2.1 make it clear that despite its consistency with the differential equation, 1.2.2 is totally useless, as the error actually *increases* as h decreases. The approximate solution is very unstable, eventually causing overflow when $h = 0.01$. Methods like 1.2.2 that are consistent but not convergent are called "unstable" methods. There are many such methods, and yet for many others—the "stable" methods—there is a close relationship between truncation error and error.

It is much easier to investigate the relationship between truncation error (or consistency) and error and the role of "stability" in this relationship, if we limit our study to the linear problem

$$\frac{du}{dt} = a(t)u + f(t), \qquad u(0) = u_0.$$

There is very little to be gained, qualitatively, by studying the general nonlinear problem, and the results will be essentially the same.

The Euler approximation to this linear problem satisfies $(t_k \equiv kh)$,

$$\frac{U(t_{k+1}) - U(t_k)}{h} = a(t_k)U(t_k) + f(t_k), \qquad U(0) = u_0.$$

Table 1.2.1
Error Growth for Unstable Method

$h = 0.02$		$h = 0.01$	
t	Error	t	Error
0.00	(exact starting value)	0.00	(exact starting value)
0.02	(exact starting value)	0.01	(exact starting value)
0.04	(exact starting value)	0.02	(exact starting value)
0.06	-0.0000000052	0.03	-0.0000000001
0.08	-0.0000000054	0.04	-0.0000000001
0.10	-0.0000000328	0.05	-0.0000000005
0.12	-0.0000000055	0.06	-0.0000000001
0.14	-0.0000001481	0.07	-0.0000000023
0.16	0.0000001379	0.08	0.0000000022
0.18	-0.0000007597	0.09	-0.0000000119
0.20	0.0000014831	0.10	0.0000000232
0.22	-0.0000047486	0.11	-0.0000000743
0.24	0.0000117214	0.12	0.0000001836
0.26	-0.0000327934	0.13	-0.0000005139
0.28	0.0000862778	0.14	0.0000013539
0.30	-0.0002335685	0.15	-0.0000036693
0.32	0.0006236405	$\vdots$	$\vdots$
0.34	-0.0016748618	0.25	-0.0713659502
0.36	0.0044836767	0.26	0.1916942484
0.38	-0.0120112118	0.27	-0.5139532162
0.40	0.0321518307	0.28	1.3847586275
0.42	-0.0859563311	0.29	-3.6789744976
0.44	0.2302225667	0.30	10.1544280920
0.46	-0.6117149090	0.31	-25.0719314973
0.48	1.6573653507	0.32	81.5127579564
0.50	-4.2333011119	0.33	-74.6693268179
		0.34	479.1500454882
		0.35	3702.3113597741
		0.36	283789.8538851262
		$\vdots$	$\vdots$

By definition of the truncation error, the exact solution u satisfies

$$\frac{u(t_{k+1}) - u(t_k)}{h} = a(t_k)u(t_k) + f(t_k) + T(\xi_k), \qquad u(0) = u_0,$$

where $T(\xi_k) = \frac{1}{2}hu''(\xi_k)$ is the truncation error. Subtracting these two equations yields a difference equation for the error, $e(t) \equiv U(t) - u(t)$:

$$\frac{e(t_{k+1}) - e(t_k)}{h} = a(t_k)e(t_k) - T(\xi_k), \qquad e(0) = 0. \tag{1.2.3}$$

Now the relationship between truncation error and error becomes clearer. If the truncation error were exactly zero, the error would satisfy a homogeneous linear difference equation with homogeneous initial condition, and the error would be identically zero for all t_k. As $h \to 0$, consistency guarantees that the nonhomogeneous term, the truncation error, approaches zero. So we would expect that the error also goes to zero—but does it? To ask this question is to ask if the difference method is *stable*:

Definition An approximate (finite difference) method is *stable* if the error goes to zero as the truncation error goes to zero.

In most textbooks, stability is defined in terms of what happens to the approximate solution as $t \to \infty$, with h fixed. These equivalent definitions, while more easily verifiable, obscure the real significance of stability.

To verify that the Euler approximation is stable, we first need the following result, which will be referenced frequently later.

Theorem 1.2.1 *If a sequence of nonnegative numbers e_k satisfies*

$$e_{k+1} \leq (1 + hA)e_k + hT,$$

where h, A, and T are nonnegative, then

$$e_k \leq \exp(At_k)[e_0 + t_k T],$$

where $t_k = kh$.

Proof

$$
\begin{aligned}
e_{k+1} &\leq (1 + hA)e_k + hT \\
&\leq (1 + hA)[(1 + hA)e_{k-1} + hT] + hT \\
&= (1 + hA)^2 e_{k-1} + [(1 + hA) + 1]hT \\
&\quad\vdots \\
&\leq (1 + hA)^{k+1} e_0 + [(1 + hA)^k + \cdots + (1 + hA) + 1]hT \\
&\leq (1 + hA)^{k+1}[e_0 + (k + 1)hT] \\
&\leq \exp[hA(k + 1)][e_0 + t_{k+1}T] \\
&= \exp[At_{k+1}][e_0 + t_{k+1}T],
\end{aligned}
$$

from which the result follows. Here the fact that $\exp(\alpha) = 1 + \alpha + \frac{1}{2}\alpha^2 + \cdots \geq 1 + \alpha$ (for nonnegative α) has been used. ∎

Now, if $T_{\max} \equiv \max |T(t)|$, $A_{\max} \equiv \max |a(t)|$, where the maxima are over the time interval of interest, then 1.2.3 implies

$$e(t_{k+1}) = [1 + a(t_k)h]e(t_k) - hT(\xi_k),$$

$$|e(t_{k+1})| \leq (1 + hA_{\max})|e(t_k)| + hT_{\max}.$$

Since $e(t_0) = 0$, applying Theorem 1.2.1 to this inequality gives

$$|e(t_k)| \leq \exp(A_{\max}t_k)t_k T_{\max}. \tag{1.2.4}$$

Thus as $T_{\max}$ goes to zero, so does the error, at a fixed value of t. In fact, since (assuming u'' is bounded) $T_{\max} = O(h)$, the bound 1.2.4 shows that the error itself is also $O(h)$, as confirmed by the experimental results on the problem 1.1.2.

For the unstable method 1.2.2, on the other hand, as $h \rightarrow 0$ the truncation error goes to zero, but the error does not, as will be shown theoretically in the next section, and as was vividly demonstrated experimentally in Table 1.2.1.

In summary, a finite difference approximation to a differential equation is consistent if the truncation error goes to zero with h. The error (in the linear case) will satisfy a difference equation that has the truncation error as a nonhomogeneous term. The difference method is called stable if the fact that the truncation error goes to zero guarantees that the solution of the difference equation—the error—also goes to zero. In other words, consistency plus stability equals convergence.

While we are discussing the topics of consistency and truncation error, it seems appropriate to derive a couple of other approximations that will be used extensively throughout the later chapters of the text.

A "central difference" approximation to $u'(t)$ is found by subtracting the Taylor series expansions

$$u(t + h) = u(t) + u'(t)h + u''(t)\frac{h^2}{2} + u'''(\xi_1)\frac{h^3}{6} \qquad (t \leq \xi_1 \leq t + h),$$

$$u(t - h) = u(t) - u'(t)h + u''(t)\frac{h^2}{2} - u'''(\xi_2)\frac{h^3}{6} \qquad (t - h \leq \xi_2 \leq t),$$

and by dividing by $2h$:

$$\frac{u(t + h) - u(t - h)}{2h} = u'(t) + \frac{h^2}{6}\left[\frac{1}{2}u'''(\xi_1) + \frac{1}{2}u'''(\xi_2)\right].$$

Now (assuming u''' is continuous) there must be a point ξ between ξ_1 and ξ_2 such that

$$\frac{u(t+h) - u(t-h)}{2h} = u'(t) + \frac{h^2 u'''(\xi)}{6} \qquad (t - h \le \xi \le t + h). \quad (1.2.5a)$$

A central difference approximation to $u''(t)$ that will be used extensively is found by adding the Taylor series expansions

$$u(t+h) = u(t) + u'(t)h + u''(t)\frac{h^2}{2} + u'''(t)\frac{h^3}{6} + u^{iv}(\xi_1)\frac{h^4}{24},$$

$$u(t-h) = u(t) - u'(t)h + u''(t)\frac{h^2}{2} - u'''(t)\frac{h^3}{6} + u^{iv}(\xi_2)\frac{h^4}{24},$$

and subtracting $2u(t)$:

$$\frac{u(t+h) - 2u(t) + u(t-h)}{h^2} = u''(t) + \frac{h^2 u^{iv}(\xi)}{12} \qquad (t - h \le \xi \le t + h).$$
$$(1.2.5b)$$

For completeness, we repeat the Euler approximation to $u'(t)$, since it will also be used frequently:

$$\frac{u(t+h) - u(t)}{h} = u'(t) + \frac{1}{2}hu''(\xi) \qquad (t \le \xi \le t + h). \quad (1.2.5c)$$

1.3   Multistep Methods

The Euler and unstable methods considered in the previous sections are special cases of the more general (m-step) multistep method

$$\frac{[U(t_{k+1}) + \alpha_1 U(t_k) + \alpha_2 U(t_{k-1}) + \cdots + \alpha_m U(t_{k+1-m})]}{h}$$

$$= \beta_0 f(t_{k+1}, U(t_{k+1})) + \cdots + \beta_m f(t_{k+1-m}, U(t_{k+1-m})). \quad (1.3.1)$$

The following generalization of Theorem 1.2.1 will be useful in studying the stability of these multistep methods:

Theorem 1.3.1   *If a sequence of numbers e_k satisfies*

$$e_{k+1} + \rho_1 e_k + \rho_2 e_{k-1} + \cdots + \rho_m e_{k+1-m} = hT_k \qquad (1.3.2)$$

for $k \geq m - 1$ $(m \geq 1)$, and if all the roots of the corresponding characteristic polynomial

$$\lambda^m + \rho_1 \lambda^{m-1} + \cdots + \rho_m \qquad (1.3.3)$$

are less than or equal to 1 in absolute value, and all multiple roots are strictly less than 1 in absolute value, then

$$|e_k| \leq M_\rho [\max\{|e_0|, \ldots, |e_{m-1}|\} + t_k T],$$

where $t_k = kh$, $T = \max |T_j|$, and M_ρ is a constant depending only on the ρ_i.

Proof The equation satisfied by e_k can be written in the form

$$
\begin{bmatrix} e_{k+1} \\ e_k \\ e_{k-1} \\ \vdots \\ e_{k-m+2} \end{bmatrix}
=
\begin{bmatrix}
-\rho_1 & -\rho_2 & -\rho_3 & \cdots & -\rho_{m-1} & -\rho_m \\
1 & 0 & 0 & \cdots & 0 & 0 \\
0 & 1 & 0 & \cdots & 0 & 0 \\
\vdots & \vdots & \vdots & & \vdots & \vdots \\
0 & 0 & 0 & \cdots & 1 & 0
\end{bmatrix}
\begin{bmatrix} e_k \\ e_{k-1} \\ e_{k-2} \\ \vdots \\ e_{k-m+1} \end{bmatrix}
+
\begin{bmatrix} hT_k \\ 0 \\ 0 \\ \vdots \\ 0 \end{bmatrix}
$$

This matrix-vector equation can be abbreviated

$$\mathbf{E_{k+1}} = A\mathbf{E_k} + \mathbf{c_k}. \qquad (1.3.4)$$

Now, $\mathbf{E_{k+1}}$ can be related back to $\mathbf{E_{m-1}}$ as follows:

$$\mathbf{E_{k+1}} = A[A\mathbf{E_{k-1}} + \mathbf{c_{k-1}}] + \mathbf{c_k} = A^2 \mathbf{E_{k-1}} + A\mathbf{c_{k-1}} + \mathbf{c_k},$$

$$\vdots$$

$$\mathbf{E_{k+1}} = A^{k-m+2} \mathbf{E_{m-1}} + A^{k-m+1} \mathbf{c_{m-1}} + A^{k-m} \mathbf{c_m} + \cdots + \mathbf{c_k},$$

$$
\begin{aligned}
\|\mathbf{E_{k+1}}\|_\infty &\leq \|A^{k-m+2}\|_\infty \|\mathbf{E_{m-1}}\|_\infty \\
&\quad + [\|A^{k-m+1}\|_\infty + \cdots + \|I\|_\infty] \max[\|\mathbf{c_{m-1}}\|_\infty, \ldots, \|\mathbf{c_k}\|_\infty].
\end{aligned}
$$

$$(1.3.5)$$

(Recall that the infinity norms for vectors and matrices, respectively, are defined by

$$\|\mathbf{x}\|_\infty \equiv \max_j |x_j|, \qquad \|A\|_\infty \equiv \max_i \sum_j |A_{ij}|$$

and that

$$\|A\mathbf{x}\|_\infty = \max_i \left| \sum_j A_{ij} x_j \right| \leq \max_i \sum_j |A_{ij}| |x_j| \leq \|A\|_\infty \|\mathbf{x}\|_\infty.)$$

There is an upper bound on the powers of the m by m matrix A, say $\|A^k\|_\infty \le M_\rho$. One way to see that this is true is to observe that the eigenvalues of A are the roots of the characteristic polynomial 1.3.3 (see Problem 2); and by a well-known result of linear algebra (proven by looking at the Jordan canonical form of A), the assumptions made about the eigenvalues (the roots) assure that there is an upper bound to the powers of A. A more direct approach is to consider 1.3.2 in the case that $T_k \equiv 0$. This mth order linear, homogeneous, constant coefficient recurrence relation has m independent solutions, one of the form $e_k = \lambda^k$ for each simple root λ of the characteristic polynomial, and l for each root of multiplicity l, namely λ^k, $k\lambda^k, \ldots, k^{l-1}\lambda^k$. The general solution of the homogeneous version of 1.3.2 is an arbitrary linear combination of these solutions, so clearly if all roots are less than or equal to 1 in absolute value and all multiple roots are less than 1 in absolute value, the homogeneous solution is bounded for all k, for any choice of starting values $e_0, e_1, \ldots, e_{m-1}$. Since 1.3.4 is equivalent to 1.3.2, however, this means that $\mathbf{E}_{k+1}$ is bounded for all k, for any starting vector $\mathbf{E}_{m-1}$, when $\mathbf{c}_k \equiv 0$. But since $\mathbf{E}_{k+1} = A^{k-m+2}\mathbf{E}_{m-1}$ in the homogeneous case, this can only be true if the powers of A are bounded.

In any case, (1.3.5) now gives

$$|e_{k+1}| \le \|\mathbf{E}_{k+1}\|_\infty \le M_\rho \max\{|e_0|\ldots|e_{m-1}|\} + M_\rho(k-m+2)hT$$
$$\le M_\rho[\max\{|e_0|\ldots|e_{m-1}|\} + t_{k+1}T],$$

which is equivalent to the desired bound. ∎

Now we consider the stability of the multistep method 1.3.1 applied to the constant coefficient linear problem $u' = au + f(t)$. For this problem, 1.3.1 becomes

$$\frac{[U(t_{k+1}) + \sum_{i=1}^{m}\alpha_i U(t_{k+1-i})]}{h} = \sum_{i=0}^{m} \beta_i[aU(t_{k+1-i}) + f(t_{k+1-i})].$$

By the definition of the truncation error, the exact solution $u(t)$ of the differential equation satisfies

$$\frac{[u(t_{k+1}) + \sum_{i=1}^{m}\alpha_i u(t_{k+1-i})]}{h} = \sum_{i=0}^{m} \beta_i[au(t_{k+1-i}) + f(t_{k+1-i})] + T(\xi_k),$$

where $T(\xi_k)$ is the truncation error, assumed to be at least $O(h)$. Subtracting these two equations yields a difference equation for the error $e(t) \equiv U(t) - u(t)$:

$$\frac{[e(t_{k+1}) + \sum_{i=1}^{m} \alpha_i e(t_{k+1-i})]}{h} = \sum_{i=0}^{m} \beta_i a e(t_{k+1-i}) - T(\xi_k)$$

or

$$e(t_{k+1}) + \sum_{i=1}^{m} \left[\frac{\alpha_i - \beta_i ah}{1 - \beta_0 ah} \right] e(t_{k+1-i}) = \frac{-hT(\xi_k)}{1 - \beta_0 ah}. \qquad (1.3.6)$$

Before stating a stability theorem, let us here define a stronger form of stability, which will be important to the development of Section 1.5.

Definition The region of *absolute stability* of a multistep method consists of those values of ah in the complex plane for which all roots of the polynomial

$$(1 - \beta_0 ah)\lambda^m + (\alpha_1 - \beta_1 ah)\lambda^{m-1} + \cdots + (\alpha_m - \beta_m ah) \qquad (1.3.7)$$

are less than or equal to 1 in absolute value, and all multiple roots are strictly less than 1 in absolute value.

A straightforward application of Theorem 1.3.1 to the recurrence relation 1.3.6, with $\rho_i = (\alpha_i - \beta_i ah)/(1 - \beta_0 ah)$ and $T_k = -T(\xi_k)/(1 - \beta_0 ah)$, shows that if ah belongs to the region of absolute stability of the multistep method, then the error is bounded by

$$|e(t_k)| \le M_\rho[\max\{|e(t_0)|, \ldots, |e(t_{m-1})|\} + t_k T_{max}/|1 - \beta_0 ah|] \qquad (1.3.8)$$

where $T_{max} \equiv \max |T(\xi_j)|$.

Now the criterion for stability of a multistep method will be stated, and an outline of the proof will be given.

Theorem 1.3.2 *The multistep method 1.3.1 is stable provided all roots of* $\lambda^m + \alpha_1 \lambda^{m-1} + \cdots + \alpha_m$ *are less than or equal to 1 in absolute value, and all multiple roots are strictly less than 1 in absolute value.*

The proof follows closely the proof of Theorem 1.3.1, with

$$\rho_i = \frac{\alpha_i - \beta_i ah}{1 - \beta_0 ah} = \alpha_i + O(ah).$$

The "A" matrix is now equal to $A_\alpha + O(ah)$, where A_α, the "A" matrix corresponding to $\rho_i = \alpha_i$, has eigenvalues (the roots of $\lambda^m + \alpha_1 \lambda^{m-1} + \cdots + \alpha_m$)

that are all less than or equal to 1 in absolute value. Since ah is not assumed to be in the absolute stability region, the eigenvalues of A are now not necessarily less than 1 in absolute value, but are bounded—it can be shown—by $1 + C_{\alpha\beta}|a|h$, where $C_{\alpha\beta} \geq 0$ is a constant. Therefore the eigenvalues of the matrix $A/(1 + C_{\alpha\beta}|a|h)$ are less than or equal to 1 in absolute value, and (at least for h below some threshold value, since these eigenvalues are converging to those of the A_α matrix) all multiple eigenvalues are less than 1 in absolute value. Hence the powers of $A/(1 + C_{\alpha\beta}|a|h)$ are bounded in norm by some constant $M_{\alpha\beta}$, and the norm of A^k (and lower powers) is bounded by $M_{\alpha\beta}(1 + C_{\alpha\beta}|a|h)^k \leq M_{\alpha\beta} \exp[C_{\alpha\beta}|a|t_k]$. Therefore the bound ($M_\rho$) on the powers of A in 1.3.8 must be replaced by this exponential term, and the error bound becomes

$$|e(t_k)| \leq M_{\alpha\beta} \exp[C_{\alpha\beta}|a|t_k]\left[\max\{|e(t_0)|, \ldots, |e(t_{m-1})|\} + \frac{t_k T_{\max}}{|1 - \beta_0 ah|}\right]. \quad (1.3.9)$$

A similar bound can be derived for the general linear problem, but this requires much more work.

An m-step method requires m initial values to get started, so the approximate solution at $t_0, \ldots, t_{m-1}$ must be calculated using another method. Formula 1.3.9 states that if the errors committed by this starting method go to zero as $h \to 0$ and if the multistep method 1.3.1 is consistent ($T_{\max} \to 0$ as $h \to 0$), the error at a fixed point t_k goes to zero. In other words, the method is stable. ■

Formula 1.3.9 also suggests that if the multistep method has truncation error of order $O(h^n)$ then, to avoid a loss of accuracy, the method used to calculate the starting values should be at least of order $O(h^{n-1})$, so that the errors $e(t_0), \ldots, e(t_{m-1})$ will be $O(h^n)$.

Now it can be seen why the method 1.2.2 is unstable. Its characteristic polynomial is

$$2\lambda^3 + 3\lambda^2 - 6\lambda + 1 = (\lambda - 1)(2\lambda^2 + 5\lambda - 1),$$

which has roots 1.0, 0.186, and -2.686. It is clearly the root -2.686 that is responsible for the instability of this method. In fact, closer examination of Table 1.2.1 shows that the error actually grows by a factor of about -2.686 each time step, once this term begins to dominate, and before the nonlinear term u^2 in the differential equation obscures this factor.

In Sections 1.4 and 1.5, specific multistep methods are given that are stable and consistent with $u' = f(t, u)$, with truncation errors of high order. According to 1.3.9 this guarantees that the errors are of high order in h.

1.4   Adams Multistep Methods

Some of the most popular high-order, stable, multistep methods are the *Adams* methods, which ensure stability by choosing $\alpha_1 = -1$ and $\alpha_2 = \cdots = \alpha_m = 0$. The characteristic polynomial corresponding to Theorem 1.3.2 is $\lambda^m - \lambda^{m-1}$, which has 1 as a simple root and 0 as a multiple root. Thus these methods are stable regardless of the values chosen for the β's.

The values of the β's in 1.3.1 are determined in order to maximize the order of the truncation error. Consider, for example, the two-step Adams method ($m = 2$):

$$\frac{U(t_{k+1}) - U(t_k)}{h} = \beta_0 f(t_{k+1}, U(t_{k+1}))$$

$$+ \beta_1 f(t_k, U(t_k)) + \beta_2 f(t_{k-1}, U(t_{k-1})). \qquad (1.4.1)$$

The truncation error is the amount by which the solution of $u' = f(t, u)$ fails to satisfy the difference equation

$$T = \frac{u(t_{k+1}) - u(t_k)}{h} - \beta_0 f(t_{k+1}, u(t_{k+1}))$$

$$- \beta_1 f(t_k, u(t_k)) - \beta_2 f(t_{k-1}, u(t_{k-1}))$$

$$= \frac{u(t_{k+1}) - u(t_k)}{h} - \beta_0 u'(t_{k+1}) - \beta_1 u'(t_k) - \beta_2 u'(t_{k-1}).$$

Expanding everything in a Taylor series about t_k gives

$$T = \frac{\left[u(t_k) + u'(t_k)h + u''(t_k)\frac{h^2}{2} + u'''(t_k)\frac{h^3}{6} + u^{iv}(\xi_1)\frac{h^4}{24} - u(t_k) \right]}{h}$$

$$- \beta_1 u'(t_k) - \beta_0 \left[u'(t_k) + u''(t_k)h + u'''(t_k)\frac{h^2}{2} + u^{iv}(\xi_2)\frac{h^3}{6} \right]$$

$$- \beta_2 \left[u'(t_k) - u''(t_k)h + u'''(t_k)\frac{h^2}{2} - u^{iv}(\xi_3)\frac{h^3}{6} \right]$$

$$= u'(t_k)[1 - \beta_0 - \beta_1 - \beta_2] + \tfrac{1}{2}hu''(t_k)[1 - 2\beta_0 + 2\beta_2]$$

$$+ \frac{1}{6}h^2 u'''(t_k)[1 - 3\beta_0 - 3\beta_2]$$

$$+ \frac{1}{24}h^3[u^{iv}(\xi_1) - 4\beta_0 u^{iv}(\xi_2) + 4\beta_2 u^{iv}(\xi_3)].$$

To maximize the order, β_0, β_1, β_2 are chosen so that

$$1 - \beta_0 - \beta_1 - \beta_2 = 0,$$
$$1 - 2\beta_0 \quad\quad + 2\beta_2 = 0,$$
$$1 - 3\beta_0 \quad\quad - 3\beta_2 = 0.$$

The unique solution to this system of linear equations is $\beta_0 = \frac{5}{12}$, $\beta_1 = \frac{8}{12}$, $\beta_2 = -\frac{1}{12}$. Then

$$T = \frac{-1}{24} h^3 \left[-u^{iv}(\xi_1) + \frac{5}{3} u^{iv}(\xi_2) + \frac{1}{3} u^{iv}(\xi_3) \right].$$

For small h, the three points ξ_1, ξ_2, and ξ_3 are closely bunched, and thus, since $-1 + \frac{5}{3} + \frac{1}{3} = 1$, the quantity in brackets almost reduces to the value of u^{iv} at a single point. In fact, it is shown later that

$$T = \frac{-1}{24} h^3 u^{iv}(\xi) \qquad (t_{k-1} \le \xi \le t_{k+1}). \qquad (1.4.2)$$

Another way the coefficients of the Adams methods can be determined is to set the right-hand side of $u'(t) = f(t, u(t))$ equal to its (Lagrange) polynomial interpolant plus the error:

$$u'(t) = \frac{(t - t_k)(t - t_{k-1})}{2h^2} f(t_{k+1}, u(t_{k+1}))$$

$$+ \frac{(t - t_{k+1})(t - t_{k-1})}{-h^2} f(t_k, u(t_k))$$

$$+ \frac{(t - t_{k+1})(t - t_k)}{2h^2} f(t_{k-1}, u(t_{k-1}))$$

$$+ \frac{(t - t_{k+1})(t - t_k)(t - t_{k-1})}{6} \frac{d^3}{dt^3} f(\xi, u(\xi)).$$

It is easily verified that the first three terms constitute a quadratic polynomial that interpolates to the function $g(t) \equiv f(t, u(t))$ at the points t_{k+1}, t_k, and t_{k-1}. The last term is the polynomial interpolation error, given in any elementary numerical analysis text. Since $u^{iv}(t) = d^3/dt^3(f(t, u(t)))$, integrating the above equation from t_k to t_{k+1} and dividing by h (to put it in normalized form) gives

$$\frac{u(t_{k+1}) - u(t_k)}{h} = \frac{5}{12} f(t_{k+1}, u(t_{k+1})) + \frac{8}{12} f(t_k, u(t_k)) - \frac{1}{12} f(t_{k-1}, u(t_{k-1}))$$

$$+ \frac{1}{6h} \int_{t_k}^{t_{k+1}} (t - t_{k+1})(t - t_k)(t - t_{k-1})u^{iv}(\xi(t)) \, dt.$$

The last term is clearly the truncation error, and, since $(t - t_{k+1})(t - t_k)$ $(t - t_{k-1})$ does not change sign in the interval of integration, there must be a point ξ in (t_{k-1}, t_{k+1}) such that

$$T = \frac{u^{iv}(\xi)}{6h} \int_{t_k}^{t_{k+1}} (t - t_{k+1})(t - t_k)(t - t_{k-1}) \, dt = \frac{-u^{iv}(\xi)h^3}{24},$$

and so 1.4.2 is proven.

A table of the coefficients β_i in 1.3.1 that maximize the truncation error order for two other values of m, along with the corresponding truncation errors, is given in Table 1.4.1. Recall that $\alpha_1 = -1, \alpha_2 = \cdots = \alpha_m = 0$ for all Adams methods. The methods in Table 1.4.1 are sometimes called Adams–Moulton methods, or corrector formulas. Because these methods are stable, the error is of the same order as the truncation error.

Table 1.4.1
Implicit Adams (Corrector) Coefficients

m	β_0	β_1	β_2	β_3	T
1	$\frac{1}{2}$	$\frac{1}{2}$			$-\frac{1}{12} \ (h^2 u'''(\xi))$
2	$\frac{5}{12}$	$\frac{8}{12}$	$-\frac{1}{12}$		$-\frac{1}{24} \ (h^3 u^{iv}(\xi))$
3	$\frac{9}{24}$	$\frac{19}{24}$	$-\frac{5}{24}$	$\frac{1}{24}$	$-\frac{19}{720} \ (h^4 u^v(\xi))$

There is an obvious problem with the Adams–Moulton, or corrector, methods such as 1.4.1; they cannot be explicitly solved for the unknown $U(t_{k+1})$ in terms of the known (previous) values of U. However, the value of $U(t_{k+1})$ that satisfies the nonlinear corrector formula can be found using the iteration

$$U^{n+1}(t_{k+1}) = U(t_k) + h\beta_0 f(t_{k+1}, U^n(t_{k+1}))$$

$$+ h\beta_1 f(t_k, U(t_k)) + \cdots + h\beta_m f(t_{k+1-m}, U(t_{k+1-m})).$$

$$\tag{1.4.3}$$

If $U(t_{k+1})$ is the exact solution of the nonlinear corrector formula, subtracting the equation satisfied by $U(t_{k+1})$ from 1.4.3 gives

$$U^{n+1}(t_{k+1}) - U(t_{k+1}) = h\beta_0 f(t_{k+1}, U^n(t_{k+1})) - h\beta_0 f(t_{k+1}, U(t_{k+1})),$$

$$|U^{n+1}(t_{k+1}) - U(t_{k+1})| \leq [h|\beta_0| \cdot |f_u(t_{k+1}, \xi^n)|] |U^n(t_{k+1}) - U(t_{k+1})|,$$

$$(1.4.4)$$

where ξ^n is between $U^n(t_{k+1})$ and $U(t_{k+1})$. If f_u is bounded near the exact solution, for sufficiently small h the term in square brackets is less than 1, and the iteration will converge to $U(t_{k+1})$. Of course we do not want to do too many iterations, so a good starting value for the iteration is needed. In fact, it has been shown that if the accuracy of the starting value is of the same order in h as the corrector formula itself, one iteration is sufficient to preserve the asymptotic accuracy of the corrector.

The usual way to get such a starting value is to use another Adams method that *can* be solved explicitly for $U(t_{k+1})$, that is, one with $\beta_0 = 0$. The Adams–Bashforth, or predictor, methods are often used. These are Adams methods (i.e., $\alpha_1 = -1, \alpha_2 = \cdots = \alpha_m = 0$) that are designed to maximize the order of the truncation error under the restriction that $\beta_0 = 0$. The coefficients β_i for four values of m are given in Table 1.4.2 along with the corresponding truncation errors. Note that since there is one less free parameter, the orders are diminished by 1 compared to the corresponding corrector formulas. These predictor formulas may be used as stand-alone methods, or they may be used each step only to generate a starting value for one of the corrector iterations 1.4.3.

The often used fourth-order predictor–corrector pair algorithm consists of using the fourth-order predictor to calculate an approximate $U(t_{k+1})$ and using this to start the iteration on the fourth-order corrector, doing only one iteration:

$$U*(t_{k+1}) = U(t_k) + \frac{55h}{24} f(t_k, U(t_k)) - \frac{59h}{24} f(t_{k-1}, U(t_{k-1}))$$

$$+ \frac{37h}{24} f(t_{k-2}, U(t_{k-2})) - \frac{9h}{24} f(t_{k-3}, U(t_{k-3}))$$

$$U(t_{k+1}) = U(t_k) + \frac{9h}{24} f(t_{k+1}, U*(t_{k+1})) + \frac{19h}{24} f(t_k, U(t_k))$$

$$- \frac{5h}{24} f(t_{k-1}, U(t_{k-1})) + \frac{h}{24} f(t_{k-2}, U(t_{k-2})). \quad (1.4.5)$$

Table 1.4.2
Explicit Adams (Predictor) Coefficients

m	β_0	β_1	β_2	β_3	β_4	T
1	0	1				$\frac{1}{2}(hu''(\xi))$
2	0	$\frac{3}{2}$	$-\frac{1}{2}$			$\frac{5}{12}(h^2u'''(\xi))$
3	0	$\frac{23}{12}$	$-\frac{16}{12}$	$\frac{5}{12}$		$\frac{3}{8}(h^3u^{iv}(\xi))$
4	0	$\frac{55}{24}$	$-\frac{59}{24}$	$\frac{37}{24}$	$-\frac{9}{24}$	$\frac{251}{720}(h^4u^v(\xi))$

Algorithm 1.4.5 was applied to the problem $u' = 2t(1 + u^2)$, $u(0) = 0$ (cf. 1.1.2). The required starting values $U(t_0)$, $U(t_1)$, $U(t_2)$, $U(t_3)$ were given exact values, so that all errors could be attributed to the predictor–corrector method. For more general problems, the starting values may be calculated using a single-step method of appropriate order, usually a Runge–Kutta method (see Section 1.6), or a Taylor series method (see Problem 7). Another popular technique is to begin with a first-order method and a very small stepsize, and gradually to increase the stepsize and order.

Results with $h = 0.02$ and $h = 0.01$ are shown in Table 1.4.3. Note that as h is decreased by a factor of 2, the error at $t = 0.5$ decreases by a factor of 15, suggesting a rate of convergence of $\ln(15)/\ln(2) = 3.9$, close to the predicted value of 4.

When the fourth order predictor alone was used, with $h = 0.01$, the error at $t = 0.5$ was $0.137E - 6$, about 13 times larger than using the predictor–corrector combined formula 1.4.5. This reflects the fact that the predictor truncation error is larger than that of the corrector by a factor of $\frac{251}{19} = 13.2$, and explains why a predictor–corrector pair is often preferred over the predictor by itself. The reason the corrector truncation errors are smaller

Table 1.4.3
Predictor–Corrector Errors

	t	Error
$h = 0.02$	0.10	$0.26E - 08$
	0.20	$0.16E - 07$
	0.30	$0.42E - 07$
	0.40	$0.84E - 07$
	0.50	$0.16E - 06$
$h = 0.01$	0.10	$0.30E - 09$
	0.20	$0.12E - 08$
	0.30	$0.28E - 08$
	0.40	$0.56E - 08$
	0.50	$0.11E - 07$

than those of the predictor is intuitively explained by noting that the predictor formulas can be derived by interpolating $f(t, u(t))$ at $t_k, t_{k-1}, \ldots$ and integrating this interpolatory polynomial outside its range of interpolation, from t_k to t_{k+1}. But in this *extrapolated* range it is less accurate than the polynomial that interpolates to $f(t, u(t))$ at $t_{k+1}, t_k, \ldots$, used to derive the corrector.

When the corrector was iterated to convergence, rather than iterated a single time, the results changed very little. For $h = 0.01$, the error at $t = 0.5$ actually increased about 2%.

A computer method that hopes to be successful on a wide range of initial value problems must be "adaptive," that is, it must be able to vary the stepsize as the calculations progress, guided by some estimate of the error. If the solution varies rapidly in some portions of the time interval of interest and slowly in others, a constant time step is inappropriate.

Unfortunately, it is very difficult to obtain a good estimate of the actual error at a given time, since the error accumulates with time in a nontrivial way. What is usually done is to monitor the "local" error—the new error introduced each step, assuming the approximate solution to be exact up to that point. That is, if u_k is the solution of the differential equation with initial condition $u_k(t_k) = U(t_k)$, the stepsize is varied to maintain the local error $U(t_{k+1}) - u_k(t_{k+1})$ within some specified tolerance. How this tolerance is related to the final, or "global," error is not clear, but at least we are controlling some reasonable measure of the error.

From 1.3.6 we see that this "new," or "local," error is (for small h) just equal to $-h$ times the truncation error, where the truncation error is defined in terms of u_k instead of u. For an Adams predictor–corrector pair, the local error can be estimated from Tables 1.4.1 and 1.4.2. For the fourth-order pair (1.4.5), we have

$$U_*(t_{k+1}) - u_k(t_{k+1}) \approx -h T_{\text{pred}}(\xi_1) = \frac{-251}{720} h^5 u_k^v(\xi_1),$$

$$U(t_{k+1}) - u_k(t_{k+1}) \approx -h T_{\text{corr}}(\xi_2) = \frac{19}{720} h^5 u_k^v(\xi_2).$$

Assuming h is small enough so that $\xi_1 \approx \xi_2$, we deduce that

$$U(t_{k+1}) - u_k(t_{k+1}) \approx \frac{19}{270} [U(t_{k+1}) - U_*(t_{k+1})].$$

and so the right-hand side can be used to estimate the local error.

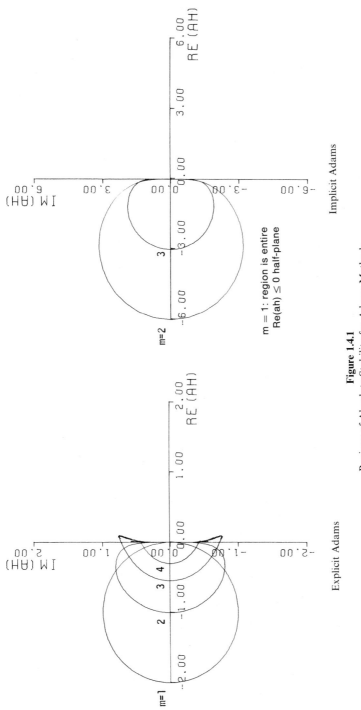

Figure 1.4.1

Regions of Absolute Stability for Adams Methods
(Methods are Absolutely Stable *Inside* Closed Curves)

It is more difficult to modify the stepsize for a multistep method than for a single-step method. Most multistep codes maintain the stepsize constant as long as feasible; when it must be changed, the required restarting values are obtained by interpolation (of appropriate degree) at the desired time points, which do not necessarily correspond to values where the solution has already been calculated.

For future reference, the regions of absolute stability for the explicit and implicit Adams methods are shown in Figure 1.4.1. Recall that the significance of absolute stability is that when $u' = au + f(t)$ is solved, the error bound does not include the potentially huge exponential term $\exp[C_{\alpha\beta}|a|t_k]$ (1.3.8 holds) when ah lies within the region of absolute stability.

If ah is a point on the boundary of the region of absolute stability, (1.3.7) must have a root of absolute value equal to one. Thus if we set $\lambda = \exp(I\theta)$ ($I = \sqrt{-1}$) in 1.3.7 and solve for ah when this polynomial is set to zero, varying θ from 0 to 2π will generate the absolute stability region boundary, plus possibly some additional curves that must be eliminated by further inspection. This was done to generate the regions shown in Figure 1.4.1.

1.5   Backward Difference Methods for Stiff Problems

The error bound 1.3.9 for a multistep method applied to the linear problem $u' = au + f(t)$ has the form $M_{\alpha\beta} \exp[C_{\alpha\beta}|a|t_k]t_k T_{\max}/|1 - \beta_0 ah|$, assuming exact starting values are used. The term $t_k T_{\max}/|1 - \beta_0 ah|$ is the error that would be expected if the global (total) error were just the sum of the "local" errors, where the local error is defined to be the error in $U(t_{k+1})$ assuming $U(t_0), \ldots, U(t_k)$ to be exact. To see this, observe that, if the previous errors are all zero, 1.3.6 shows that the error at t_{k+1} is $e(t_{k+1}) = -hT(\xi_k)/(1 - \beta_0 ah)$. Thus the sum of the local errors is bounded by

$$\left| \sum_{i=1}^{k} \frac{-hT(\xi_i)}{(1 - \beta_0 ah)} \right| \leq \frac{t_k T_{\max}}{|1 - \beta_0 ah|}.$$

The term $\exp[C_{\alpha\beta}|a|t_k]$, on the other hand, reflects the fact that the influence of early local errors may grow exponentially with time. If $|a|$ is large, this exponential term may be extremely large, and while 1.3.9 ensures that if h is sufficiently small the error can be made arbitrarily small, it may require an unreasonably small value of h to bring the error down to an acceptable level. When $|a|$ is large, stability is sufficient only in theory: In practice what is needed is absolute stability (see Section 1.3), which ensures that the error bound 1.3.8 (which does not include the exponential term) holds.

It is instructive to display the analytic solution of the constant coefficient linear problem $u' = au + f(t)$, $u(0) = u_0$:

$$u(t) = u_0 e^{at} + \int_0^t e^{a(t-s)} f(s) \, ds.$$

From this it can be seen that when a is large and positive (or $\text{Re}(a)$ is large and positive), the solution varies rapidly and has large derivatives, and even the local error is large (e.g., $T = \frac{1}{2} h u''$ for Euler's method). When $\text{Re}(a)$ is large and negative, however, the solution is very smooth and well behaved (if $f(t)$ is, also), and the local error can be made small with a reasonable value of h (see Problem 3). The problem $u' = au + f(t)$ when $\text{Re}(a)$ is large and negative is called a "stiff" differential equation. Occasionally when $\text{Re}(a)$ is large and positive, the problem is also called stiff. In either case, the multistep error bound is much larger than would be expected by looking at the local, or truncation, error. In the former case this problem is especially annoying since the solution is smooth, yet an extremely small value of h is required to obtain a reasonable error.

Since the exponential term $\exp[C_{\alpha\beta}|a|t_k]$ contains the time variable, a scale-independent definition of stiffness is as follows:

Definition The problem $u' = au + f(t)$ is called *stiff* if $\text{Re}(ah*)$ is large and negative, where $h*$ is some measure of the time scale of the problem.

Systems of first-order equations of the form $\mathbf{u}' = \mathbf{f}(t, \mathbf{u})$ exhibit this stiffness behavior when the Jacobian matrix f_u has some eigenvalues whose real parts are large and negative. (Notice that if A is diagonalizable, $A = P^{-1}DP$, the linear system $\mathbf{u}' = A\mathbf{u} + \mathbf{f}(t)$ reduces to a set of uncoupled equations $z_i' = \lambda_i z_i + q_i(t)$, where the z_i and q_i are the components of $\mathbf{z} = P\mathbf{u}$ and $\mathbf{q} = Pf$, respectively, and the λ_i are the eigenvalues of A.) The fact that these eigenvalues may be complex explains why we have allowed for the possibility that a may be complex. Such systems will be encountered when time-dependent partial differential equations such as the diffusion equation (Chapter 2) are solved. In fact, for these problems some eigenvalues will not only be large and negative, but, as the spatial discretization is refined, they will converge to $-\infty$! These applications are the main motivation for our interest in stiff problems in this text.

Since when $\text{Re}(a)$ is large and negative the solution is still smooth, it seems there ought to be a way to design finite difference methods that do not require a very small stepsize for these stiff problems, and there is. The backward difference formulas are multistep methods of the form 1.3.1 with $\beta_1 = \cdots = \beta_m = 0$, and with $\beta_0 > 0$, $\alpha_1, \ldots, \alpha_m$ chosen to maximize the order of the truncation error. Table 1.5.1 shows the values of the backward difference coefficients for several values of m.

Table 1.5.1
Backward Difference Coefficients

m	β_0	α_1	α_2	α_3	α_4	T
1	1	-1				$-\frac{1}{2}\,(hu''(\xi))$
2	$\frac{2}{3}$	$-\frac{4}{3}$	$\frac{1}{3}$			$-\frac{1}{3}\,(h^2u'''(\xi))$
3	$\frac{6}{11}$	$-\frac{18}{11}$	$\frac{9}{11}$	$-\frac{2}{11}$		$-\frac{1}{4}\,(h^3u^{iv}(\xi))$
4	$\frac{12}{25}$	$-\frac{48}{25}$	$\frac{36}{25}$	$-\frac{16}{25}$	$\frac{3}{25}$	$-\frac{1}{5}\,(h^4u^{v}(\xi))$

First let us check that these formulas are stable (and thus eventually convergent), that is, that they satisfy the root condition of Theorem 1.3.2. The characteristic polynomials and the absolute values of their roots are given below:

m	Characteristic Polynomial	Absolute Values of Roots
1	$\lambda - 1$	1
2	$3\lambda^2 - 4\lambda + 1$	1, 0.333
3	$11\lambda^3 - 18\lambda^2 + 9\lambda - 2$	1, 0.426, 0.426
4	$25\lambda^4 - 48\lambda^3 + 36\lambda^2 - 16\lambda + 3$	1, 0.561, 0.561, 0.381

The backward difference formulas in Table 1.5.1 are stable. It can be shown that for $m = 5$ and 6 the backward difference methods are also stable, but not for $m = 7$.

But we want more than stability now, we want multistep methods that are absolutely stable when $\text{Re}(ah)$ is large and negative. When such methods are applied to stiff problems ($\text{Re}(ah*)$ large and negative), h can be of the same order of magnitude as the time scale of the problem ($h*$), and the error bound 1.3.8, which contains no exponential term, holds. Then the global error is essentially just the sum of the local, or truncation, errors—the ideal situation.

The backward difference methods satisfy this requirement. To verify this, note that for these methods the characteristic equation 1.3.7 takes the form

$$(1 - \beta_0 ah)\lambda^m + \alpha_1 \lambda^{m-1} + \cdots + \alpha_m = 0. \tag{1.5.1}$$

Now, if $|ah|$ is large enough so that

$$|1 - \beta_0 ah| > \sum_{i=1}^{m} |\alpha_i|,$$

then all roots λ satisfy $|\lambda| < 1$, because if we suppose there is a root with $|\lambda| \geq 1$, then, from (1.5.1),

$$|1 - \beta_0 ah| = |-\alpha_1 \lambda^{-1} - \alpha_2 \lambda^{-2} - \cdots - \alpha_m \lambda^{-m}| \leq \sum_{i=1}^{m} |\alpha_i|,$$

contradicting the assumption.

The backward difference methods are absolutely stable not only when Re(ah) is large and negative, but for all values of ah outside some finite region. The exact regions of absolute stability, for the first four backward difference methods, are shown in Figure 1.5.1.

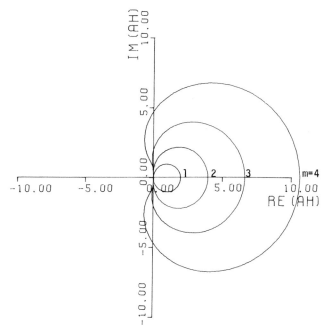

Figure 1.5.1
Regions of Absolute Stability for Backward Difference Methods
(Methods are Absolutely Stable *Outside* Closed Curves)

Unfortunately, the backward difference formulas are all implicit, that is, they cannot be solved for $U(t_{k+1})$ in terms of the known values. However, it is impossible to construct an explicit method that does *not* perform poorly on stiff systems (Problem 4). Implicitness must be considered the price to be paid for efficiency on stiff problems.

Another method useful for stiff problems is the second-order implicit Adams–Moulton method (Table 1.4.1, $m = 1$):

$$U(t_{k+1}) - U(t_k) = \tfrac{1}{2}hf(t_{k+1}, U(t_{k+1})) + \tfrac{1}{2}hf(t_k, U(t_k)).$$

For this method the characteristic polynomial 1.3.7 is

$$(1 - \tfrac{1}{2}ha)\lambda + (-1 - \tfrac{1}{2}ha),$$

which has the root $\lambda = (1 + \tfrac{1}{2}ha)/(1 - \tfrac{1}{2}ha)$.

If $\mathrm{Re}(ah)$ is negative, $|\lambda| < 1$ as is easily verified, so that this method, to be encountered later under the name of Crank–Nicolson, is well suited for stiff systems also. However, when $\mathrm{Re}(ah)$ is positive, $|\lambda| > 1$. (See Figure 1.4.1.)

To demonstrate the usefulness of the backward difference methods, we solved the stiff problem

$$u' = -1000u + \sin(t), \qquad u(0) = \frac{-1}{1000001}, \tag{1.5.2}$$

which has a smooth solution,

$$u(t) = \frac{1000\sin(t) - \cos(t)}{1000001}.$$

Three third-order methods were tried: an explicit Adams–Bashforth method (Table 1.4.2 with $m = 3$), an implicit Adams–Moulton method (Table 1.4.1 with $m = 2$), and a backward difference method (Table 1.5.1 with $m = 3$). Since this problem is linear, it is possible to explicitly solve for $U(t_{k+1})$ even for the two implicit methods, and this is what was done, so that no predictor or iteration was needed. Exact values were used to start each multistep method, and the results are shown in Table 1.5.2.

The explicit Adams method was useless until N was large enough so that $ah = -1000/N$ was within the absolute stability region ($N > 1833$; see Figure 1.4.1), when it suddenly gave good results. The implicit Adams method did much better, reflecting the fact that its absolute stability region is larger, extending out to $ah = -6$, but it had some stability problems until $-1000h$ was greater than -6 ($N > 167$). The backward difference method, in further testing, gave good results for h as large as 0.1.

Table 1.5.2

Comparison of Three Third-Order Methods on a Stiff Problem

$N(=1/h)$	Relative Error at $t = 1$		
	Explicit Adams	Implicit Adams	Backward Difference
100	OVERFLOW	0.36E − 03	0.26E − 09
200	OVERFLOW	0.55E − 11	0.33E − 10
300	OVERFLOW	0.16E − 11	0.97E − 11
400	OVERFLOW	0.68E − 12	0.41E − 11
500	OVERFLOW	0.35E − 12	0.21E − 11
600	OVERFLOW	0.20E − 12	0.12E − 11
700	OVERFLOW	0.13E − 12	0.76E − 12
800	OVERFLOW	0.86E − 13	0.51E − 12
900	OVERFLOW	0.60E − 13	0.36E − 12
1000	OVERFLOW	0.44E − 13	0.26E − 12
1100	OVERFLOW	0.33E − 13	0.20E − 12
1200	OVERFLOW	0.25E − 13	0.15E − 12
1300	OVERFLOW	0.20E − 13	0.12E − 12
1400	OVERFLOW	0.16E − 13	0.95E − 13
1500	OVERFLOW	0.13E − 13	0.78E − 13
1600	OVERFLOW	0.11E − 13	0.64E − 13
1700	0.13E + 35	0.92E -- 14	0.53E − 13
1800	0.17E − 03	0.76E − 14	0.45E − 13
1900	0.57E − 13	0.68E − 14	0.38E − 13
2000	0.49E − 13	0.56E − 14	0.33E − 13

For stiff systems, $|f_u|$ is large, so it may require a very small value of h to make a simple corrector iteration converge (see 1.4.4). But a small h is precisely what we are trying to avoid by using implicit methods, so this is generally unacceptable. A much more useful iteration is Newton's method, which applied to a general implicit multistep method 1.3.1 is

$$U^{n+1}(t_{k+1}) = U^n(t_{k+1}) - [1 - h\beta_0 f_u(t_{k+1}, U^n(t_{k+1}))]^{-1}$$

$$\times \left[U^n(t_{k+1}) + \sum_{i=1}^{m} \alpha_i U(t_{k+1-i}) - h\beta_0 f(t_{k+1}, U^n(t_{k+1})) \right.$$

$$\left. - \sum_{i=1}^{m} h\beta_i f(t_{k+1-i}, U(t_{k+1-i})) \right]. \qquad (1.5.3)$$

For a system of N differential equations, 1.5.3 is still valid, provided

$$[1 - h\beta_0 f_u(t_{k+1}, U^n(t_{k+1}))]^{-1} \equiv A^{-1}$$

is interpreted as the inverse of the identity matrix minus $h\beta_0$ times the Jacobian of f. The inverse of A is not actually formed explicitly, as this is very inefficient, but rather the LU decomposition (see Section 0.3) of A is formed. Then the calculation of $A^{-1}\mathbf{f} = U^{-1}L^{-1}\mathbf{f}$ is done by solving two triangular linear systems. If A varies slowly with time, or not at all, the LU decomposition done on one time step may be saved and used on several subsequent time steps. As long as the old A is still reasonably accurate, Newton's method will still converge (more slowly), and solving the two triangular systems is much less expensive ($O(N^2)$ work for full systems) than calculating the LU decomposition itself ($O(N^3)$ work for full systems).

State-of-the-art software for stiff initial value problems, such as Hindmarsh's implementation of Gear's backward difference methods [Hindmarsh, 1974], employs adaptive schemes for deciding when to update and redecompose A, as well as for deciding when the time step needs to be decreased or increased. Most also optionally calculate the elements of the Jacobian matrix using finite differences, so that the user does not have to supply these himself.

1.6   Runge–Kutta Methods

Probably the easiest to program of all high-order methods for initial value problems are the Runge–Kutta methods. These are one-step methods that, unlike the multistep methods, are self-starting, and we have no difficulty varying the stepsize. On the other hand, they require several evaluations of f per step.

A typical Runge–Kutta formula requiring two function evaluations per step has the form

$$v_1 = f(t_k, U(t_k)),$$

$$v_2 = f(t_k + \alpha h, U(t_k) + \alpha h v_1), \qquad (1.6.1)$$

$$U(t_{k+1}) = U(t_k) + h[Av_1 + Bv_2].$$

Let us determine values for α, A, and B that make as high-order as possible the truncation error, which is calculated by substituting the solution of $u' = f(t, u)$ into the difference equation, in normalized form:

$$T = \frac{u(t_{k+1}) - u(t_k)}{h} - Af(t_k, u(t_k)) - Bf(t_k + \alpha h, u(t_k) + \alpha h v_1)$$

$$= u'(t_k) + \tfrac{1}{2}hu''(t_k) + \tfrac{1}{6}h^2 u'''(t_k) + O(h^3)$$

$$- Au'(t_k) - Bf(t_k + \alpha h, u(t_k) + \alpha h u'(t_k)).$$

Expanding this last term by using the bivariate Taylor's series and using the relations

$$u' = f(t, u),$$

$$u'' = f_t(t, u) + f_u(t, u)u',$$

$$u''' = f_{tt}(t, u) + 2f_{tu}(t, u)u' + f_{uu}(t, u)[u']^2 + f_u(t, u)u'',$$

we get

$$T = (1 - A)u' + \tfrac{1}{2}hu'' + \tfrac{1}{6}h^2u''' + O(h^3)$$

$$- B[f + \alpha hf_t + \alpha hu'f_u + \tfrac{1}{2}\alpha^2h^2f_{tt} + \alpha^2h^2u'f_{tu} + \tfrac{1}{2}\alpha^2h^2[u']^2f_{uu} + O(h^3)]$$

$$= (1 - A)u' + \tfrac{1}{2}hu'' + \tfrac{1}{6}h^2u''' + O(h^3)$$

$$- B[u' + \alpha hu'' + \tfrac{1}{2}\alpha^2h^2[u''' - f_uu'']]$$

$$= [1 - A - B]u' + [\tfrac{1}{2} - B\alpha]hu''$$

$$+ [\tfrac{1}{6} - \tfrac{1}{2}B\alpha^2]h^2[u''' - f_uu''] + \tfrac{1}{6}h^2f_uu'' + O(h^3),$$

where u and its derivatives are evaluated at t_k, and f and its derivatives are evaluated at $(t_k, u(t_k))$.

For a general function f, it is impossible to choose A, B, α so that the $O(h^2)$ terms disappear, because of the presence of the last f_uu'' term. Therefore, we must be content with choosing these parameters so that the constant and $O(h)$ terms are zero. This means $B = 0.5/\alpha$, $A = 1 - 0.5/\alpha$, where α can be chosen arbitrarily. One such choice is $\alpha = 1$, $A = B = \tfrac{1}{2}$, in which case 1.6.1 is equivalent to the second-order Adams–Moulton corrector (Table 1.4.1 with $m = 1$) with Euler's method used as a predictor.

A popular fourth-order Runge–Kutta method is

$$v_1 = f(t_k, U(t_k)),$$

$$v_2 = f(t_k + \tfrac{1}{2}h, U(t_k) + \tfrac{1}{2}hv_1),$$

$$v_3 = f(t_k + \tfrac{1}{2}h, U(t_k) + \tfrac{1}{2}hv_2),$$

$$v_4 = f(t_k + h, U(t_k) + hv_3),$$

$$\tag{1.6.2}$$

$$U(t_{k+1}) = U(t_k) + \frac{h}{6}[v_1 + 2v_2 + 2v_3 + v_4].$$

To prove that the truncation error is fourth-order is obviously a difficult task, but it has been done. Clearly, however, v_1, v_2, v_3, and v_4 are intended to estimate $u'(t)$ at t_k, $t_k + \tfrac{1}{2}h$, $t_k + \tfrac{1}{2}h$ and $t_k + h$, respectively. The last equation in 1.6.2 can be understood as an attempt to apply Simpson's rule to estimate the integral in

$$u(t_{k+1}) - u(t_k) = \int_{t_k}^{t_{k+1}} u'(t)\, dt.$$

Because they are self-starting, Runge–Kutta methods can be used to calculate the starting values for a multistep method, or they can be used as stand-alone formulas.

Runge–Kutta methods are very easy to program, and they work quite well on nonstiff problems. Applying the fourth-order method 1.6.2 to the problem $u' = 2t(1 + u^2)$, $u(0) = 0$ (cf. 1.1.2) gave the results shown in Table 1.6.1.

Table 1.6.1
Runge–Kutta Method on
NonStiff Problem

	t	Error
$h = 0.02$	0.10	0.13E − 09
	0.20	0.53E − 09
	0.30	0.12E − 08
	0.40	0.23E − 08
	0.50	0.38E − 08
$h = 0.01$	0.10	0.10E − 10
	0.20	0.30E − 10
	0.30	0.80E − 10
	0.40	0.14E − 09
	0.50	0.24E − 09

Comparing these results with those of the fourth-order Adams predictor–corrector 1.4.5, in Table 1.4.3, it can be seen that the Runge–Kutta method has an error about 40 times smaller when the same stepsize is used on this same problem. (It should be recalled, however, that the predictor–corrector method requires fewer function evaluations per step.) Comparing the errors at $t = 0.5$ with $h = 0.02$ and $h = 0.01$, it is seen that the error decreases by a factor of 15.8 as h is halved. Thus, its fourth-order accuracy is confirmed experimentally. When 1.6.2 is applied to the *stiff* problem

$$u' = -1000u + \sin(t), \qquad u(0) = -1/1000001$$

(cf. 1.5.2), the results (Table 1.6.2) suggest that Runge–Kutta methods are not well suited for stiff problems, and, even after the fourth-order method 1.6.2

becomes stable, the error is not nearly as small as for the third-order implicit multistep methods (see Table 1.5.2).

As mentioned in Section 1.4, a computer method that hopes to be successful on a wide range of initial value problems must be adaptive and must control at least the local error (the new error introduced at each step, assuming the computations to be exact up to that point). The Runge–Kutta–Fehlberg methods estimate this local error by taking a step with an $O(h^m)$ Runge–Kutta method, giving an estimate $U(t_{k+1})$, and then repeating the step with another method of order $O(h^{m+1})$ accuracy, giving an approximation $U_{m+1}(t_{k+1})$. For small h the higher-order estimate can be taken as an approximation to $u_k(t_{k+1})$, where u_k is the solution of the differential equation with initial condition $u_k(t_k) = U(t_k)$. Then

$$U(t_{k+1}) - u_k(t_{k+1}) \approx U(t_{k+1}) - U_{m+1}(t_{k+1}),$$

so the right-hand side may be used to estimate the local error. Actually, it makes more sense to monitor the estimated local error per unit time

$$\frac{U(t_{k+1}) - U_{m+1}(t_{k+1})}{h},$$

Table 1.6.2
Runge–Kutta Method on
Stiff Problem

$N(= 1/h)$	Relative Error at $t = 1$
100	OVERFLOW
200	OVERFLOW
300	OVERFLOW
400	0.29E − 05
500	0.50E − 06
600	0.18E − 06
700	0.86E − 07
800	0.46E − 07
900	0.27E − 07
1000	0.17E − 07
1100	0.11E − 07
1200	0.74E − 08
1300	0.52E − 08
1400	0.38E − 08
1500	0.28E − 08
1600	0.21E − 08
1700	0.17E − 08
1800	0.13E − 08
1900	0.10E − 08
2000	0.83E − 09

rather than the local error, and to decrease h when this estimate is larger than a specified tolerance or increase h when the estimate is much smaller than required.

For the lower-order method, the points at which the function f is evaluated are chosen to coincide with some of the higher-order method's function evaluation points (recall that we had some flexibility in choosing the points, i.e., in choosing α, in deriving 1.6.1). For example, when $m = 4$, the Runge–Kutta–Fehlberg fifth-order method requires evaluation of f at six points, and five of these same points are the points used by the fourth-order method. In this way, the number of function evaluations is no more than would be required by the fifth-order method alone.

The formulas for this Runge–Kutta–Fehlberg algorithm are given below. Quantities that are vectors when a *system* of first-order differential equations, $\mathbf{u}' = \mathbf{f}(t, \mathbf{u})$, is solved are indicated by boldface type.

$$\mathbf{v_1} = \mathbf{f}(t_k, \mathbf{U}(t_k)),$$

$$\mathbf{v_2} = \mathbf{f}\left(t_k + \frac{1}{4} h, \mathbf{U}(t_k) + h\,\frac{\mathbf{v_1}}{4} \right),$$

$$\mathbf{v_3} = \mathbf{f}\left(t_k + \frac{3}{8} h, \mathbf{U}(t_k) + h\,\frac{3\mathbf{v_1} + 9\mathbf{v_2}}{32} \right),$$

$$\mathbf{v_4} = \mathbf{f}\left(t_k + \frac{12}{13} h, \mathbf{U}(t_k) + h\,\frac{1932\mathbf{v_1} - 7200\mathbf{v_2} + 7296\mathbf{v_3}}{2197} \right),$$

$$\mathbf{v_5} = \mathbf{f}\left(t_k + 1h, \mathbf{U}(t_k) + h\,\frac{8341\mathbf{v_1} - 32832\mathbf{v_2} + 29440\mathbf{v_3} - 845\mathbf{v_4}}{4104} \right),$$

$$\mathbf{v_6} = \mathbf{f}\Bigg(t_k + \frac{1}{2} h,$$
$$\mathbf{U}(t_k) + h\,\frac{-6080\mathbf{v_1} + 41040\mathbf{v_2} - 28352\mathbf{v_3} + 9295\mathbf{v_4} - 5643\mathbf{v_5}}{20520} \Bigg),$$

$$\mathbf{U_4}(t_{k+1}) = \mathbf{U}(t_k) + h\,\frac{2375\mathbf{v_1} + 0\mathbf{v_2} + 11264\mathbf{v_3} + 10985\mathbf{v_4} - 4104\mathbf{v_5}}{20520},$$

$$\mathbf{U_5}(t_{k+1})$$
$$= \mathbf{U}(t_k) + h\,\frac{33440\mathbf{v_1} + 0\mathbf{v_2} + 146432\mathbf{v_3} + 142805\mathbf{v_4} - 50787\mathbf{v_5} + 10260\mathbf{v_6}}{282150},$$

$$\mathbf{U}(t_{k+1}) = \mathbf{U_4}(t_{k+1}),$$

$$E \text{ (local error per unit time)} \approx \frac{\|\mathbf{U_4}(t_{k+1}) - \mathbf{U_5}(t_{k+1})\|}{h}.$$

A primitive adaptive FORTRAN77 subroutine to implement this algorithm on a general system of first-order differential equations is shown in Figure 1.6.1. When the estimated local error per unit time is larger than the user-supplied tolerance, the step is repeated with the stepsize cut in half. If this estimate is more than 100 times smaller than requested, the stepsize is doubled the next step.

```fortran
      SUBROUTINE RFK(FSUB,N,T0,TFINAL,HINIT,U,TOL)
      IMPLICIT DOUBLE PRECISION(A-H,O-Z)
C
C     ARGUMENT DESCRIPTIONS
C
C     FSUB  -(INPUT) THE NAME OF A USER-SUPPLIED SUBROUTINE TO EVALUATE
C            THE RIGHT HAND SIDE(S) OF THE DIFFERENTIAL EQUATION SYSTEM
C            U' = F(T,U). FSUB SHOULD HAVE THE CALLING SEQUENCE
C                     SUBROUTINE FSUB(N,T,U,F)
C            AND SHOULD EVALUATE F(1)...F(N), GIVEN N,T AND U(1)...U(N).
C            FSUB MUST BE MENTIONED IN AN EXTERNAL STATEMENT IN THE
C            CALLING PROGRAM.
C     N     -(INPUT) THE NUMBER OF DIFFERENTIAL EQUATIONS (N .LE. 100)
C     T0    -(INPUT) THE INITIAL VALUE OF T.
C     TFINAL-(INPUT) THE FINAL VALUE OF T.
C     HINIT -(INPUT) THE INITIAL STEPSIZE TO BE USED.
C     U     -(INPUT) A VECTOR OF LENGTH N, HOLDING THE INITIAL VALUES OF
C            U AT T=T0.
C            (OUTPUT) THE SOLUTION VALUES AT T=TFINAL.
C            THE SOLUTION COMPONENTS ARE PRINTED EACH TIME STEP.
C     TOL   -(INPUT) THE DESIRED BOUND ON THE LOCAL ERROR PER UNIT TIME
C            OF THE SOLUTION COMPONENTS.
C
      PARAMETER (NMAX=100)
      DIMENSION U(N),V1(NMAX),V2(NMAX),V3(NMAX),V4(NMAX),V5(NMAX),
     &          V6(NMAX),U4(NMAX),U5(NMAX),UTEMP(NMAX)
      TK = T0
      H = HINIT
    5 TKP1 = MIN(TK+H , TFINAL)
      H = TKP1-TK
C                              TAKE A STEP WITH STEPSIZE H
      CALL FSUB(N , TK , U , V1)
      DO 10 I=1,N
   10 UTEMP(I) = U(I) + H*V1(I)/4
      CALL FSUB(N , TK+H/4 , UTEMP , V2)
      DO 15 I=1,N
   15 UTEMP(I) = U(I) + H*(3*V1(I)+9*V2(I)) / 32
      CALL FSUB(N , TK+3*H/8 , UTEMP , V3)
      DO 20 I=1,N
   20 UTEMP(I) = U(I) + H*(1932*V1(I)-7200*V2(I)+7296*V3(I)) / 2197
      CALL FSUB(N , TK+12*H/13 , UTEMP , V4)
      DO 25 I=1,N
```

(continued)

```
   25 UTEMP(I) = U(I) + H*(8341*V1(I)-32832*V2(I)+29440*V3(I)
      &                      -845*V4(I)) / 4104
      CALL FSUB(N , TK+H , UTEMP , V5)
      DO 30 I=1,N
   30 UTEMP(I) = U(I) + H*(-6080*V1(I)+41040*V2(I)-28352*V3(I)
      &                    +9295*V4(I)-5643*V5(I)) / 20520
      CALL FSUB(N , TK+H/2 , UTEMP , V6)
      E = 0.0
      DO 35 I=1,N
         U4(I) = U(I) + H*(2375*V1(I)+11264*V3(I)+10985*V4(I)
      &                   -4104*V5(I)) / 20520
         U5(I) = U(I) + H*(33440*V1(I)+146432*V3(I)+142805*V4(I)
      &                   -50787*V5(I)+10260*V6(I)) / 282150
   35 E = MAX(E , ABS(U4(I)-U5(I))/H)
C                             IF ERROR TOO LARGE, REPEAT STEP WITH H/2
      IF (E .GT. TOL) THEN
         H = H/2
         GO TO 5
      ENDIF
C                             OTHERWISE, UPDATE TK AND UK
      TK = TKP1
      DO 40 I=1,N
   40 U(I) = U4(I)
      PRINT 45, TK, (U(I),I=1,N)
   45 FORMAT (1X,5E15.5,/,(16X,4E15.5))
C                                   IF ERROR MUCH SMALLER THAN REQUIRED, DOUBLE
C                                   VALUE OF H FOR NEXT STEP
      IF (E .LT. 0.01*TOL) H = 2*H
      IF (TK.LT.TFINAL) GO TO 5
      RETURN
      END
```

Figure 1.6.1
Primitive Adaptive Runge-Kutta-Fehlberg Subroutine

Clearly some further refinements are needed, such as a check to make sure that the stepsize is within some reasonable bounds and the ability to output the solution at points more convenient for the user, before this subroutine is ready for a subroutine library. As it is, however, this very simple routine can solve a good number of problems accurately and efficiently.

This routine was applied to solve the problem

$$u' = \frac{100}{1 + 10000t^2}, \qquad -1 \le t \le 1,$$

$$u(-1) = \text{Arctan}(-100).$$

1.6 RUNGE-KUTTA METHODS   63

The solution, $u(t) = \text{Arctan}(100t)$, is very flat in most of the interval $[-1, 1]$, but near $t = 0$ it rises rapidly from $-\frac{1}{2}\pi$ to $\frac{1}{2}\pi$. An adaptive routine would therefore be expected to choose a very small stepsize near $t = 0$ and a larger stepsize elsewhere. This is precisely what our subroutine does, as can be seen from results in Table 1.6.3, which were obtained using TOL $= 10^{-2}$ and HINIT $= 0.25$.

Table 1.6.3
Adaptive RKF Subroutine Results

t_k	$U(t_k)$	True Error	$h_k = t_k - t_{k-1}$
−0.7500	−1.557464	0.161E − 06	0.25000
−0.2500	−1.531609	0.792E − 03	0.50000
−0.1250	−1.491882	0.916E − 03	0.12500
−0.0625	−1.413283	0.114E − 02	0.06250
−0.0313	−1.262548	0.145E − 02	0.03125
−0.0156	−1.003067	0.158E − 02	0.01563
−0.0078	−0.664734	0.153E − 02	0.00781
−0.0000	−0.001822	0.182E − 02	0.00781
0.0078	0.661661	0.154E − 02	0.00781
0.0156	0.999909	0.157E − 02	0.00781
0.0234	1.165940	0.157E − 02	0.00781
0.0313	1.259519	0.157E − 02	0.00781
0.0469	1.359044	0.157E − 02	0.01563
0.0625	1.410572	0.157E − 02	0.01563
0.0938	1.462966	0.157E − 02	0.03125
0.1563	1.505332	0.155E − 02	0.06250
0.2813	1.533724	0.153E − 02	0.12500
0.5313	1.550461	0.151E − 02	0.25000
1.0000	1.559293	0.150E − 02	0.46875

The average time step taken by the adaptive method was about 0.107. When the problem was re-solved with a *constant* stepsize of 0.107, the error at $t = 1$ was 2.79. This is about 2000 times larger than for the adaptive method, which took the same number of steps with the same Runge-Kutta formulas.

The adaptive program of Figure 1.6.1 was also applied to the stiff problem 1.5.2. Although the solution $u(t) = [1000 \sin(t) - \cos(t)]/1000001$ is very smooth, a very small stepsize was taken—about 0.003, regardless of the error tolerance supplied. The small stepsize was clearly required for stability rather than accuracy considerations (see Table 1.6.2), and when an explicit adaptive method appears to be taking a stepsize much smaller than would seem to be required for accurate approximation, stiffness of the differential equation being solved should be suspected (see Problem 8).

1.7   Problems

1. Check the following finite difference approximations to $u' = f(t, u)$ for consistency and stability. For the consistent approximations, give the truncation error.

    a. $$\frac{8U(t_{k+1}) - 9U(t_k) + U(t_{k-2})}{6h}$$

    $$= \frac{1}{2} f(t_{k+1}, U(t_{k+1})) + f(t_k, U(t_k)) - \frac{1}{2} f(t_{k-1}, U(t_{k-1})).$$

    b. $$\frac{U(t_{k+1}) - U(t_{k-2})}{3h} = \frac{1}{2} f(t_k, U(t_k)) + \frac{1}{2} f(t_{k-1}, U(t_{k-1})).$$

    c. $$\frac{U(t_{k+1}) + 4U(t_k) - 5U(t_{k-1})}{6h} = \frac{2}{3} f(t_k, U(t_k)) + \frac{1}{3} f(t_{k-1}, U(t_{k-1})).$$

    d. $$\frac{U(t_{k+1}) - U(t_k)}{h} = f(t_k, U(t_k)) + f(t_{k-1}, U(t_{k-1})).$$

2. Show that the eigenvalues of the matrix A appearing in 1.3.4 are the roots of the characteristic polynomial 1.3.3.
    Hint: Expand $\det(A - \lambda I)$ along the first row.

3. a. Show that the solution to the constant coefficient problem

    $$u' = au + f(t), \qquad 0 \le t \le t_f,$$

    $$u(0) = u_0$$

    is given by

    $$u(t) = u_0 e^{at} + \int_0^t e^{a(t-s)} f(s) \, ds.$$

    b. If $\text{Re}(a)$ is large and negative, show that, after the rapidly decaying transient terms have been damped out, $u(t)$ is approximately equal to $-f(t)/a$.

4. Show that the region of absolute stability for any *explicit* multistep method is bounded (cf. Figure 1.4.1), and hence no explicit multistep method can hope to perform well on stiff systems.
    Hint: Assuming $\beta_0 = 0$, show that when ah is in the region of absolute stability, the coefficient $\alpha_1 - \beta_1 ah$ of λ^{m-1} in the characteristic polynomial

1.3.7 is bounded in absolute value by $m!/[l!(m-l)!]$. Since at least one of the β_l is nonzero, $|ah|$ is bounded.

5. Calculate the truncation error for the Runge–Kutta method 1.6.2 in the case that f is a function of t only, and verify that it is fourth-order, at least for the problem $u' = f(t)$.

6. Write a FORTRAN subroutine that solves a system of first-order initial value differential equations, $\mathbf{u}' = \mathbf{f}(t, \mathbf{u})$, $\mathbf{u}(t_0) = \mathbf{u_0}$, using the fourth-order Runge–Kutta method 1.6.2 with a constant, user-supplied, stepsize, H. Use the subroutine RKF of Figure 1.6.1 as a starting point for the design of your subroutine. Use this subroutine to solve:
   a. The second-order problem

$$u'' = 3u' - 2u + t,$$

$$u(0) = 1.75,$$

$$u'(0) = 1.5,$$

   after it has been reduced to a system of two first-order equations. The exact solution is $u(t) = e^t + 0.5t + 0.75$.
   b. The system

$$u' = -500u + 6889v, \qquad u(0) = 83,$$

$$v' = 36u - 500v, \qquad v(0) = 6.$$

   The exact solution is $u(t) = 83e^{-2t}$, $v(t) = 6e^{-2t}$. This is a stiff system (why?), so a small stepsize will be required to avoid instability.

7. Solve the problem 1.1.2:

$$u' = 2t(1 + u^2), \qquad u(0) = 0,$$

   using the third-order Adams–Bashforth formula (Table 1.4.2, $m = 3$), with starting values $U(h)$ and $U(2h)$ calculated from
   a. A second-order Runge–Kutta formula (1.6.1 with $\alpha = 1$, $A = B = \frac{1}{2}$).
   b. The Taylor series expansion $u(kh) \approx u(0) + u'(0)kh$, where $u(0) = 0$, and $u'(0)$ is calculated by substituting $t = 0$ into the differential equation. (This is equivalent to Euler's method.)
   c. The Taylor series expansion $u(kh) \approx u(0) + u'(0)kh + \frac{1}{2}u''(0)k^2h^2$, where $u(0)$ and $u'(0)$ are as above, and $u''(0)$ is calculated by differentiating both sides of the differential equation and then substituting $t = 0$.

In each case, calculate the error at $t = 1$ (the exact solution is $u(t) = \tan(t^2)$) for two different small values of h and compute the experimental rate of convergence. Use formula 1.3.9 to explain the observed rates.

8. a. Solve the stiff system of Problem 6b by using IMSL subroutine IVPRK, which uses a fifth- and sixth-order Runge–Kutta pair, or a similar routine from your computer center library if IMSL is not available. The stiffness of this problem will force IVPRK to take steps much smaller than the smoothness of the solution would seem to require.

   b. Solve the stiff system of Problem 6b using IMSL subroutine IVPAG, with PARAM(12) $\equiv$ METH = 2. This routine uses a backward difference method and should perform much better on stiff problems.

9. Consider the ordinary differential equations system $\mathbf{u}' = A\mathbf{u}$, where A is a diagonalizable matrix with all eigenvalues real and negative. Extending our definition of stiffness in an obvious way to encompass ordinary differential equations systems, show that this system should be considered to be stiff if $|\lambda_{max}/\lambda_{min}|$ is large, where λ_{max} and λ_{min} are the largest and smallest eigenvalues of A, in absolute value.

   Hint: Show that after the more rapidly decaying solution components have been damped out, a "measure of the time scale of the problem" can be taken to be $h^* = 1/|\lambda_{min}|$.

   This is how the idea of stiffness is introduced in many textbooks, but it incorrectly implies that only systems of ordinary differential equations can be stiff.

2

The Initial Value
Diffusion Problem

2.0   Introduction

Suppose $u(x, y, z, t)$ represents the density (mass/volume) of a diffusing substance at the point (x, y, z) and at time t. We want to derive, and eventually learn how to solve, a partial differential equation satisfied by u.

To do this we need first to study an elusive vector quantity called the "flux" of u, which is best defined by $\mathbf{J} = u\mathbf{v}_{ave}$, where $\mathbf{v}_{ave}$ is the average velocity of the molecules or atoms of the diffusing substance at (x, y, z, t). We can think of $\mathbf{v}_{ave}$ as being calculated by

$$\mathbf{v}_{ave} = \frac{\sum m_i \mathbf{v_i}}{\sum m_i},$$

where the sum is over all particles with mass m_i and velocity $\mathbf{v_i}$, in a small volume. If a small window of cross-sectional area dA is placed near (x, y, z) at time t, then the total mass of substance u passing through this window during the small time interval dt will be approximately $u\mathbf{n}^T\mathbf{v}_{ave} \, dt \, dA$, where $\mathbf{n}$ is the unit normal to the window in the direction in which the flow of u is measured. This is derived by considering that every diffusing particle in the region has

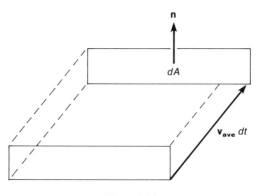

Figure 2.0.1

velocity $\mathbf{v}_{\text{ave}}$, and thus any particle within the parallelopiped whose base is the window and whose third edge is the vector $\mathbf{v}_{\text{ave}} \, dt$ (Figure 2.0.1) will pass through the window. The total mass in this parallelopiped is the density, u, times the volume, $\mathbf{n}^T \mathbf{v}_{\text{ave}} \, dt \, dA$, or $u\mathbf{n}^T \mathbf{v}_{\text{ave}} \, dt \, dA$. The mass per unit time passing through the window is

$$u\mathbf{n}^T \mathbf{v}_{\text{ave}} \, dA = \mathbf{J}^T \mathbf{n} \, dA.$$

Now let us consider an arbitrary small volume E in the region of interest. The total mass per unit time entering through the boundary ∂E of E is the sum of the mass passing through all of the boundary windows:

$$\iint_{\partial E} -\mathbf{J}^T \mathbf{n} \, dA.$$

The minus sign is present because "entering" means in the direction of $-\mathbf{n}$, with $\mathbf{n}$ being the outward unit normal. If the substance u is also being created at the rate $f(x, y, z, t)$ (mass per unit volume, per unit time) due to sources (sinks when f is negative), then

$$\frac{d}{dt}\left[\iiint_E u \, dV \right] = \iint_{\partial E} -\mathbf{J}^T \mathbf{n} \, dA + \iiint_E f \, dV.$$

This equation states that the rate of change of the total mass in the volume E is equal to the rate at which it enters through the boundary of E plus the rate at which it is generated within E. Applying the divergence theorem (which is just the multivariate generalization of the fundamental theorem of calculus) to the surface integral gives

$$\iiint_E u_t\, dV = \iiint_E \left[-\mathbf{V}^T \mathbf{J} + f \right] dV.$$

Since E is an arbitrary volume, it can be made so small that the integrands are almost constant, so that at any point

$$u_t = -\mathbf{V}^T \mathbf{J} + f. \qquad (2.0.1)$$

This is the partial differential equation that models diffusion. The relationship between the flux and u in the case of isotropic (direction independent) diffusion is known experimentally to be

$$\mathbf{J} = -D\,\mathbf{V}\mathbf{u},$$

which states that the flux (or average velocity) is in the direction of the negative of the gradient of u, i.e., in the direction of most rapid decrease in concentration. The magnitude of the flux is also proportional to the magnitude of this gradient and to a proportionality constant, D, called the diffusion coefficient.

In the case of a liquid or gas medium, there may also be present a current or wind with velocity $\mathbf{v}$. In this case, the flux of u is due not only to diffusion but also to convection by the current or wind. Then the convective flux is given by $u\mathbf{v}$ (recall that flux is density times average velocity), so that the total flux (diffusion plus convection) is

$$\mathbf{J} = -D\,\mathbf{V}\mathbf{u} + u\mathbf{v}. \qquad (2.0.2)$$

In the general case of isotropic diffusion and convection, 2.0.1 and 2.0.2 combine to give

$$u_t = \mathbf{V}^T(D\,\mathbf{V}\mathbf{u}) - \mathbf{V}^T(u\mathbf{v}) + f \qquad (2.0.3)$$

or, if $\mathbf{v} = (v_1, v_2, v_3)$,

$$u_t = (Du_x)_x + (Du_y)_y + (Du_z)_z - (uv_1)_x - (uv_2)_y - (uv_3)_z + f,$$

where v_1, v_2, v_3, D, and f may depend on x, y, z, t, and even u.

As for all differential equations, some auxiliary conditions must be specified. Usually the density u at all points in the region is specified at $t = 0$, and on each part of the boundary, at all times, either the density or the boundary flux, $\mathbf{J}^T\mathbf{n}$, is given (see Problem 1). Thus the problem has initial and boundary conditions, but it will soon be clear that qualitatively it has the character of an initial value, not a boundary value, problem.

It should be noted here that heat conduction is really just diffusion of heat energy, so that the equation 2.0.3 is valid for the heat energy $u = c\rho T$, where c, ρ, and T, respectively, are the specific heat, density, and temperature. If c and ρ are constant, a partial differential equation for the temperature is

$$c\rho T_t = \mathbf{V}^T(K\,\mathbf{V}T) - c\rho\mathbf{V}^T(T\mathbf{v}) + f. \tag{2.0.4}$$

Here $K = Dc\rho$ is called the conductivity, and f is the heat generation rate, due to heat sources and sinks.

Many of the qualitative properties of the solutions to the diffusion, or heat, equation are most easily seen by studying

$$u_t = D\,\mathbf{V}^2u,$$

$$u(x, y, z, 0) = h(x, y, z), \qquad -\infty < x < \infty,\ -\infty < y < \infty,\ -\infty < z < \infty. \tag{2.0.5}$$

These properties are of interest because numerical methods should be designed in such a way as to yield approximate solutions that share these properties as far as is possible.

Now 2.0.5 can be solved analytically, and the result is

$$u(x, y, z, t) = \int_{-\infty}^{\infty}\int_{-\infty}^{\infty}\int_{-\infty}^{\infty} h(\alpha, \beta, \Gamma)g(x - \alpha, y - \beta, z - \Gamma, t)\,d\alpha\,d\beta\,d\Gamma, \tag{2.0.6}$$

where

$$g(x - \alpha, y - \beta, z - \Gamma, t)$$

$$= (4\pi Dt)^{-3/2}\exp\left[\frac{-(\alpha - x)^2 - (\beta - y)^2 - (\Gamma - z)^2}{4Dt}\right].$$

Note that g is always positive and that its integral is equal to 1 (see Problem 2) so that the solution at a point (x, y, z) at a time $t > 0$ is just a weighted average of the initial values. This average is weighted most heavily toward points near (x, y, z), especially when t is small, but note that the solution at (x, y, z, t) for any $t > 0$ depends on the initial values *throughout all of* $\mathbf{R}^3$. In other words, the diffusion or conduction speed is infinite (this is not quite true physically, of course).

Another consequence of the fact that $u(x, y, z, t)$ is a weighted average of the initial values is that the maximum and minimum solution values at a given $t > 0$ lie within the extremes of the initial data. Also, it can be seen that when $t > 0$, all derivatives of u are continuous, no matter how discontinuous h and its derivatives may be, since all the derivatives of u are integrals of integrable functions. All of this is quite reasonable physically, since diffusion or heat conduction exercises a smoothing and averaging effect, and this is still true even if convection or source terms are also present in the partial differential equation, or if different boundary conditions are imposed. If the diffusion is negligibly slow compared to the convection, however, the nature of the problem changes dramatically (see Section 3.0).

2.1   An Explicit Method

If the solution u to the diffusion equation is, because of some symmetry, a function only of time and one space variable, the problem 2.0.3 can, in the linear case, be put into the general form

$$u_t = D(x, t)u_{xx} - v(x, t)u_x + a(x, t)u + f(x, t),$$

$$u(0, t) = g_0(t), \qquad 0 \leq t \leq t_f,$$

$$u(1, t) = g_1(t), \tag{2.1.1}$$

$$u(x, 0) = h(x), \qquad 0 \leq x \leq 1.$$

The diffusion coefficient D is assumed to be positive.

The auxiliary conditions specify initial and boundary conditions on u. An example showing how to handle more general boundary conditions will be presented in Section 2.3.

For later reference, we define

$$D_{min} = \min D(x, t), \qquad D_{max} = \max D(x, t),$$
$$V_{max} = \max |v(x, t)|, \qquad A_{max} = \max |a(x, t)|, \tag{2.1.2}$$

where the minimum and maxima are taken over $0 \leq x \leq 1, 0 \leq t \leq t_f$.

To approximate the solution of 2.1.1, we first mark off a rectangular grid in the (x, t) plane, as shown in Figure 2.1.1, where $x_i = i\Delta x$ ($\Delta x = 1/N$) and $t_k = k\,\Delta t$. The approximation to u will be denoted by U. There are numerous finite difference formulas to choose from to approximate the partial derivatives in 2.1.1. We will use the simple Euler (forward difference) approximation

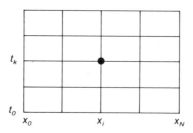

Figure 2.1.1

1.2.5c to approximate u_t (with $x = x_i$ held constant) and the central differences 1.2.5a–b to approximate u_x and u_{xx} (with $t = t_k$ held constant).

$$\frac{U(x_i, t_{k+1}) - U(x_i, t_k)}{\Delta t}$$

$$= D(x_i, t_k) \frac{U(x_{i+1}, t_k) - 2U(x_i, t_k) + U(x_{i-1}, t_k)}{\Delta x^2}$$

$$- v(x_i, t_k) \frac{U(x_{i+1}, t_k) - U(x_{i-1}, t_k)}{2\,\Delta x}$$

$$+ a(x_i, t_k)U(x_i, t_k) + f(x_i, t_k) \qquad (2.1.3)$$

with

$$U(x_0, t_k) = g_0(t_k),$$

$$U(x_N, t_k) = g_1(t_k),$$

$$U(x_i, t_0) = h(x_i).$$

The initial and boundary conditions in 2.1.3 can be used to specify the values of U at the circled points in Figure 2.1.2. Then the equation 2.1.3 can be

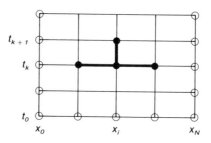

Figure 2.1.2

solved explicitly for $U(x_i, t_{k+1})$ in terms of values of U at the previous time level:

$$U(x_i, t_{k+1}) = \left[1 - \frac{2D(x_i, t_k)\,\Delta t}{\Delta x^2} + a(x_i, t_k)\,\Delta t \right] U(x_i, t_k)$$

$$+ \left[\frac{D(x_i, t_k)\,\Delta t}{\Delta x^2} + \frac{\frac{1}{2}v(x_i, t_k)\,\Delta t}{\Delta x} \right] U(x_{i-1}, t_k)$$

$$+ \left[\frac{D(x_i, t_k)\,\Delta t}{\Delta x^2} - \frac{\frac{1}{2}v(x_i, t_k)\,\Delta t}{\Delta x} \right] U(x_{i+1}, t_k)$$

$$+ \Delta t\, f(x_i, t_k). \tag{2.1.4}$$

This formula can be used to calculate the unknown values in Figure 2.1.2; first $U(x_i, t_1)$, $i = 1, \ldots, N-1$, then $U(x_i, t_2)$, etc. The "stencil" representing this explicit method, showing which unknowns are related by 2.1.4, is shown in dark lines.

Therefore, the initial value partial differential equation 2.1.1 can be solved by marching forward in time, starting with the given initial values, as was done in Chapter 1 for initial value ordinary differential equations.

The truncation error for 2.1.3 is, as before, the amount by which the exact solution fails to satisfy the finite difference equation. This can be calculated using formulas 1.2.5a–c, provided it is remembered that u is a function of both x and t so that, for example, in formula 1.2.5c it is understood that $x = x_i$ throughout. Thus,

$$T_i^k = \tfrac{1}{2}\Delta t\, u_{tt}(x_i, \xi_{ik}) - \tfrac{1}{12}D(x_i, t_k)\,\Delta x^2 u_{xxxx}(\alpha_{ik}, t_k)$$

$$+ \tfrac{1}{6}v(x_i, t_k)\,\Delta x^2 u_{xxx}(\beta_{ik}, t_k),$$

which is $O(\Delta t) + O(\Delta x^2)$, assuming the indicated derivatives are bounded, and as Δt and Δx go to zero, the finite difference equations better and better approximate the partial differential equation, i.e., the numerical method is consistent with the partial differential equation.

As discussed in Chapter 1, however, consistency does not guarantee convergence for initial value problems—the method must also be stable. To study the stability of this explicit method, let us first find the difference equation satisfied by the error. The exact solution, by the definition of truncation error, satisfies a finite difference equation almost identical to 2.1.3 but with $f(x_i, t_k)$ replaced by $f(x_i, t_k) + T_i^k$. The exact solution also satisfies the initial and boundary conditions in 2.1.3, so the error $e(x_i, t_k) \equiv$

$U(x_i, t_k) - u(x_i, t_k)$ satisfies an equation determined by subtracting the equations satisfied by U and u:

$$\frac{e(x_i, t_{k+1}) - e(x_i, t_k)}{\Delta t}$$

$$= D(x_i, t_k) \frac{e(x_{i+1}, t_k) - 2e(x_i, t_k) + e(x_{i-1}, t_k)}{\Delta x^2}$$

$$- v(x_i, t_k) \frac{e(x_{i+1}, t_k) - e(x_{i-1}, t_k)}{2\,\Delta x} + a(x_i, t_k)e(x_i, t_k) - T_i^k$$

with

$$e(x_0, t_k) = 0,$$
$$e(x_N, t_k) = 0,$$
$$e(x_i, t_0) = 0,$$

so

$$
\begin{aligned}
e(x_i, t_{k+1}) = {} & [1 - 2D(x_i, t_k)\,\Delta t/\Delta x^2 + a(x_i, t_k)\,\Delta t]e(x_i, t_k) \\
& + [D(x_i, t_k)\,\Delta t/\Delta x^2 + \tfrac{1}{2}v(x_i, t_k)\,\Delta t/\Delta x]e(x_{i-1}, t_k) \\
& + [D(x_i, t_k)\,\Delta t/\Delta x^2 - \tfrac{1}{2}v(x_i, t_k)\,\Delta t/\Delta x]e(x_{i+1}, t_k) \\
& - \Delta t\, T_i^k.
\end{aligned}
\tag{2.1.5}
$$

Because of the consistency of this explicit method, the truncation error goes to zero as Δx and Δt go to zero. It would appear from 2.1.5 that as $T_i^k \to 0$ the error would go to zero, but, as we saw in Chapter 1, the method is by definition stable if and only if this is true.

To bound the error by using 2.1.5, we will first assume (see 2.1.2) that

$$\Delta x\, V_{\max} \leq 2D_{\min},$$

so that the coefficients of $e(x_{i-1}, t_k)$ and $e(x_{i+1}, t_k)$ are nonnegative (recall that D is positive). This is certainly satisfied eventually as Δt and Δx go to zero, although if the convection speed $V_{\max}$ is very large relative to $D_{\min}$, it may require an excessively small Δx to ensure this.

We define $T_{\max} \equiv \max |T_i^k|$ as before and

$$E_k \equiv \max_{0 \leq i \leq N} |e(x_i, t_k)|.$$

Then from 2.1.5 we get

$$|e(x_i, t_{k+1})| \le |1 - 2D(x_i, t_k)\,\Delta t/\Delta x^2|E_k$$

$$+ |D(x_i, t_k)\,\Delta t/\Delta x^2 + \tfrac{1}{2}v(x_i, t_k)\,\Delta t/\Delta x|E_k$$

$$+ |D(x_i, t_k)\,\Delta t/\Delta x^2 - \tfrac{1}{2}v(x_i, t_k)\,\Delta t/\Delta x|E_k$$

$$+ A_{max}\,\Delta t\,E_k + \Delta t\,T_{max}. \qquad (2.1.6)$$

If we make the additional assumption that

$$\Delta t \le \frac{\tfrac{1}{2}\Delta x^2}{D_{max}}, \qquad (2.1.7)$$

then the first three terms on the right under absolute value signs in (2.1.6) are nonnegative. Thus the absolute value signs may be removed, and

$$|e(x_i, t_{k+1})| \le E_k + A_{max}\,\Delta t\,E_k + \Delta t\,T_{max} \qquad (i = 1, \ldots, N-1).$$

Since $e(x_0, t_{k+1}) = e(x_N, t_{k+1}) = 0$,

$$E_{k+1} \le (1 + A_{max}\,\Delta t)E_k + \Delta t\,T_{max}.$$

Theorem 1.2.1 can now be applied to this, and since $E_0 = 0$,

$$E_k \le \exp(A_{max}t_k)t_k\,T_{max}.$$

The maximum error at a fixed time t_k goes to zero as $T_{max} \to 0$, and 2.1.3 is stable, under the assumption that Δt and Δx go to zero with relative speeds such that $\Delta t \le \tfrac{1}{2}\,\Delta x^2/D_{max}$.

Note that the values of v, a, and f do not affect the stability criterion for the linear problem 2.1.1. In fact, it is generally true that only the higher-order terms influence stability. Intuitively, this is because the approximations to the higher-order derivatives are divided by higher powers of Δx and thus have larger coefficients.

The explicit method 2.1.3 can be thought of as resulting from the application of Euler's method to the system of ordinary differential equations

$$U_i'(t) = D(x_i, t)\frac{U_{i+1}(t) - 2U_i(t) + U_{i-1}(t)}{\Delta x^2} \qquad (i = 1, \ldots, N-1)$$

$$- v(x_i, t)\frac{U_{i+1}(t) - U_{i-1}(t)}{2\,\Delta x}$$

$$+ a(x_i, t)U_i(t) + f(x_i, t), \qquad (2.1.8)$$

where $U_i(t)$ represents $U(x_i, t)$, and $U_0(t)$, $U_N(t)$ are defined by the boundary conditions.

It is shown in Problem 3 that if 2.1.8 is written in vector form $\mathbf{U}' = B\mathbf{U} + \mathbf{b}$, the (algebraically) smallest eigenvalue of B is about $-4D/\Delta x^2$ (if D is constant), and that this is a stiff ordinary differential equation system that becomes more and more stiff as $\Delta x \to 0$. It is therefore to be expected that Euler's method will give bad results until Δt is small enough that $(-4D/\Delta x^2)\,\Delta t$ is within the region of absolute stability for Euler's method, that is, until $(-4D/\Delta x^2)\,\Delta t \geq -2$. (Euler's method is the first-order explicit Adams method—see Figure 1.4.1.) This is equivalent to the condition 2.1.7 found earlier, when D is constant.

2.2 Implicit Methods

Although Euler's method and other explicit multistep methods may be used to solve the ordinary differential equation system 2.1.8, they will all require excessively small time steps for stability because of the stiffness of the system. Thus we are led to try some of the multistep methods discussed in Section 1.5, which are well designed for stiff systems. These are all implicit, which means a system of (linear or nonlinear, as the partial differential equation is linear or nonlinear) algebraic equations must be solved each time step. For the one-dimensional problem this is not a serious drawback, since the linear systems involved are tridiagonal.

If we apply, for example, the first-order backward difference method (Table 1.5.1, $m = 1$) to 2.1.8, we get the implicit method:

$$\frac{U(x_i, t_{k+1}) - U(x_i, t_k)}{\Delta t}$$

$$= D(x_i, t_{k+1})\frac{U(x_{i+1}, t_{k+1}) - 2U(x_i, t_{k+1}) + U(x_{i-1}, t_{k+1})}{\Delta x^2}$$

$$- v(x_i, t_{k+1})\frac{U(x_{i+1}, t_{k+1}) - U(x_{i-1}, t_{k+1})}{2\,\Delta x}$$

$$+ a(x_i, t_{k+1})U(x_i, t_{k+1}) + f(x_i, t_{k+1})$$

or

$$\left[1 + \frac{2D(x_i, t_{k+1})\,\Delta t}{\Delta x^2} - a(x_i, t_{k+1})\,\Delta t\right] U(x_i, t_{k+1})$$

$$+ \left[\frac{-D(x_i, t_{k+1})\,\Delta t}{\Delta x^2} - \frac{\frac{1}{2}v(x_i, t_{k+1})\,\Delta t}{\Delta x}\right] U(x_{i-1}, t_{k+1})$$

$$+ \left[\frac{-D(x_i, t_{k+1})\,\Delta t}{\Delta x^2} + \frac{\frac{1}{2}v(x_i, t_{k+1})\,\Delta t}{\Delta x}\right] U(x_{i+1}, t_{k+1})$$

$$= U(x_i, t_k) + \Delta t\, f(x_i, t_{k+1}) \qquad (i = 1, \ldots, N-1). \qquad (2.2.1)$$

Each of these $N-1$ equations involves three unknowns, so a tridiagonal linear system must be solved at each time step for the $N-1$ unknown values $U(x_i, t_{k+1})$, $i = 1, \ldots, N-1$. Figure 2.2.1 shows the stencil corresponding to the implicit method 2.2.1. Values known from the initial and boundary conditions are circled.

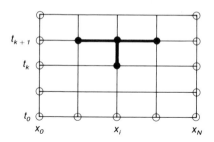

Figure 2.2.1

If it is assumed that Δt and Δx are small enough that (see 2.1.2)

$$\Delta x\, V_{\max} \le 2D_{\min},$$

$$\Delta t\, A_{\max} < 1, \qquad (2.2.2)$$

then this tridiagonal system is diagonal dominant, that is, the diagonal entry in each row (the coefficient of $U(x_i, t_{k+1})$) is larger in absolute value than the sum of the absolute values of the off-diagonal terms (see Section 0.2). For, under the assumptions 2.2.2, the two off-diagonal coefficients are nonpositive, so

$$\left| \frac{-D(x_i, t_{k+1})\,\Delta t}{\Delta x^2} - \frac{\frac{1}{2}v(x_i, t_{k+1})\,\Delta t}{\Delta x} \right| + \left| \frac{-D(x_i, t_{k+1})\,\Delta t}{\Delta x^2} + \frac{\frac{1}{2}v(x_i, t_{k+1})\,\Delta t}{\Delta x} \right|$$

$$= \frac{D(x_i, t_{k+1})\Delta t}{\Delta x^2} + \frac{\frac{1}{2}v(x_i, t_{k+1})\,\Delta t}{\Delta x} + \frac{D(x_i, t_{k+1})\,\Delta t}{\Delta x^2} - \frac{\frac{1}{2}v(x_i, t_{k+1})\,\Delta t}{\Delta x}$$

$$= \frac{2D(x_i, t_{k+1})\,\Delta t}{\Delta x^2}$$

$$< \left| 1 + \frac{2D(x_i, t_{k+1})\,\Delta t}{\Delta x^2} - a(x_i, t_{k+1})\,\Delta t \right|.$$

Diagonal dominance assures not only that a unique solution of this linear system exists, but also that Gaussian elimination may be done without pivoting (see Theorem 0.2.1). An algorithm such as the one shown in Figure 0.4.4 may be used to solve the tridiagonal system.

An analysis of the stability of the implicit method 2.2.1 similar to that carried out for the explicit method leads first to an equation for the error analogous to 2.1.5:

$$[1 + 2D(x_i, t_{k+1})\,\Delta t/\Delta x^2 - a(x_i, t_{k+1})\,\Delta t]e(x_i, t_{k+1})$$

$$+ \,[-D(x_i, t_{k+1})\,\Delta t/\Delta x^2 - \tfrac{1}{2}v(x_i, t_{k+1})\,\Delta t/\Delta x]e(x_{i-1}, t_{k+1})$$

$$+ \,[-D(x_i, t_{k+1})\,\Delta t/\Delta x^2 + \tfrac{1}{2}v(x_i, t_{k+1})\,\Delta t/\Delta x]e(x_{i+1}, t_{k+1})$$

$$= e(x_i, t_k) - \Delta t\, T_i^k, \tag{2.2.3}$$

where the truncation error, T_i^k, is still $O(\Delta t) + O(\Delta x^2)$.

With the assumptions 2.2.2 still in force,

$$[1 + 2D(x_i, t_{k+1})\,\Delta t/\Delta x^2 - a(x_i, t_{k+1})\,\Delta t]|e(x_i, t_{k+1})|$$

$$\leq |D(x_i, t_{k+1})\,\Delta t/\Delta x^2 + \tfrac{1}{2}v(x_i, t_{k+1})\,\Delta t/\Delta x|E_{k+1}$$

$$+ |D(x_i, t_{k+1})\,\Delta t/\Delta x^2 - \tfrac{1}{2}v(x_i, t_{k+1})\,\Delta t/\Delta x|E_{k+1} + E_k + \Delta t\, T_{max}$$

$$\leq 2D(x_i, t_{k+1})\,(\Delta t/\Delta x^2)E_{k+1} + E_k + \Delta t\, T_{max},$$

where T_{max} and E_k are defined as before. Note that the coefficient of $e(x_i, t_{k+1})$ is positive, by assumptions 2.2.2. Then

$$|e(x_i, t_{k+1})| \leq \frac{2D(x_i, t_{k+1})(\Delta t/\Delta x^2)E_{k+1} + E_k + \Delta t\, T_{max}}{1 + 2D(x_i, t_{k+1})\,\Delta t/\Delta x^2 - a(x_i, t_{k+1})\,\Delta t}.$$

Now if $i = l$ is the index that maximizes the right-hand side over all $1 \le i \le N - 1$, then (since $e(x_i, t_{k+1}) = 0$ for $i = 0$ and N)

$$E_{k+1} \le \frac{2D(x_l, t_{k+1})(\Delta t/\Delta x^2)E_{k+1} + E_k + \Delta t\, T_{max}}{1 + 2D(x_l, t_{k+1})\, \Delta t/\Delta x^2 - a(x_l, t_{k+1})\, \Delta t}$$

and

$$E_{k+1}[1 - a(x_l, t_{k+1})\, \Delta t] \le E_k + \Delta t\, T_{max}$$

or

$$E_{k+1} \le \frac{E_k + \Delta t\, T_{max}}{1 - A_{max}\, \Delta t}.$$

Since

$$\frac{1}{1 - A_{max}\, \Delta t} = 1 + A_{max}\, \Delta t + O(\Delta t^2),$$

then, for sufficiently small Δt, the $O(\Delta t^2)$ term will be less than Δt and $1 - A_{max}\, \Delta t \ge \frac{1}{2}$. Then, for Δt below some threshold value,

$$E_{k+1} \le [1 + (A_{max} + 1)\, \Delta t]E_k + 2\, \Delta t\, T_{max}.$$

Applying Theorem 1.2.1 to this gives (since $E_0 = 0$)

$$E_k \le \exp[(A_{max} + 1)t_k] \cdot 2t_k T_{max}.$$

The only assumptions that had to be made to obtain this error bound were of the form $\Delta x \le$ constant or $\Delta t \le$ constant. So the implicit method is "unconditionally" stable, meaning the error goes to zero as Δx and Δt go to zero with any relative speeds. This contrasts with the condition $\Delta t \le \frac{1}{2}\Delta x^2/D_{max}$ (formula 2.1.7) for stability of the explicit method, which requires that Δt go to zero much faster than Δx.

All of the implicit backward difference methods of Table 1.5.1 have been used very successfully to solve ordinary differential equation systems of the form 2.1.8. Another popular method is the second-order Adams–Moulton corrector formula (Table 1.4.1, $m = 1$), which when applied to 2.1.8 is called the Crank–Nicolson method. It is shown in Appendix 1 that the Crank–Nicolson method is also unconditionally stable on 2.1.8, and it can be shown that its truncation error is $O(\Delta t^2) + O(\Delta x^2)$.

It might be wondered if it is possible to devise an explicit method that is stable for reasonably large values of the time step. The answer is no, and this

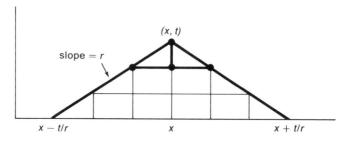

Figure 2.2.2

can most clearly be seen by considering the "domain of dependence" of an explicit method. If Δt and Δx go to zero with any fixed, nonzero, ratio $\Delta t/\Delta x = r$, the value of the approximate solution at a fixed point (x, t) will depend only on the initial values in some finite interval. For example, if U at (x_i, t_{k+1}) depends on U at (x_{i+1}, t_k), (x_i, t_k), and (x_{i-1}, t_k), as for the explicit method 2.1.3, then (see Figure 2.2.2) the approximate solution at (x, t) depends only on the values of $h(x)$ between $x - t/r$ and $x + t/r$. As Δt and Δx go to zero, with constant ratio, all the approximate solutions, and therefore their limits, depend only on initial values within the base of the triangle drawn.

But the true solution depends, as shown in Section 2.0, on *all* the values of $h(x)$. Thus the approximate solution cannot possibly be converging to the true solution, in general, because we can change the initial values outside this triangle and change the true solution but not the limit of the approximate solutions! If we use any explicit method, we must refine Δt faster than any multiple of Δx to have a chance of convergence. For an implicit method, on the other hand, the approximate solution at any point (x, t), $t > 0$ depends on all the initial values, because it is calculated by solving a system of simultaneous linear equations each step. Thus implicitness is the inevitable price for unconditional stability.

2.3   A One-Dimensional Example

Consider the one-dimensional diffusion problem

$$u_t = u_{xx} + xtu_x + xtu,$$

$$u(0, t) = e^t \qquad 0 \le t \le 1,$$

$$u_x(1, t) = -u(1, t),$$

$$u(x, 0) = e^{-x}, \qquad 0 \le x \le 1,$$

$$(2.3.1)$$

which has the exact solution $u(x, t) = e^{t-x}$. All of the explicit and implicit methods mentioned in the previous sections are equivalent to certain multistep methods applied to the ordinary differential equation system (cf. 2.1.8)

$$U_i'(t) = \frac{U_{i+1}(t) - 2U_i(t) + U_{i-1}(t)}{\Delta x^2}$$

$$+ x_i t \frac{U_{i+1}(t) - U_{i-1}(t)}{2\Delta x} + x_i t\, U_i(t), \qquad (2.3.2a)$$

$$U_i(0) = e^{-x_i} \qquad (i = 1, \ldots, N),$$

where $U_i(t)$ represents $U(x_i, t)(x_i = i\,\Delta x, \Delta x = 1/N)$.

The first boundary condition translates into $U_0(t) = e^t$. The second is unfamiliar, but can be approximated by

$$\frac{U_{N+1}(t) - U_{N-1}(t)}{2\Delta x} = -U_N(t).$$

These two equations can be used to eliminate $U_0(t)$ and $U_{N+1}(t)$ from the first and Nth ordinary differential equations in 2.3.2a. The new form for the first and Nth equations is then

$$U_1'(t) = \frac{U_2(t) - 2U_1(t) + e^t}{\Delta x^2} + x_1 t \frac{U_2(t) - e^t}{2\Delta x} + x_1 t U_1(t),$$

$$U_N'(t) = \frac{(-2\Delta x - 2)U_N(t) + 2U_{N-1}(t)}{\Delta x^2}. \qquad (2.3.2b)$$

This system of ordinary differential equations was solved using Euler's method (cf. 2.1.3) and the first- and fourth-order backward difference implicit methods (Table 1.5.1, $m = 1$ and $m = 4$), with results as shown in Table 2.3.1. For the fourth-order method, the required starting values on the time levels t_1, t_2, and t_3 were obtained from the (known) exact solution, but could have been calculated using a Taylor series expansion (using a procedure similar to that adopted in Section 3.3).

For the implicit methods, the subroutine in Figure 0.4.4 was used to solve the tridiagonal linear systems. Euler's method exhibits the behavior typical of explicit methods on stiff problems—when the time step is very small (satisfying 2.1.7), the error is small, but above the stability limit the method suddenly becomes worthless. On the other hand, the backward difference methods are useful with much larger time step sizes. For the fourth-order formula, a very small Δx was used so that all the error could be attributed to

Table 2.3.1

Comparison of Three Methods for
Diffusion Problems

Euler's Method (Explicit)

Δx	Δt	t	Maximum Error
0.0200	0.0002	0.1	0.459E − 05
		0.2	0.481E − 05
		0.3	0.486E − 05
		0.4	0.580E − 05
		0.5	0.691E − 05
		0.6	0.818E − 05
		0.7	0.964E − 05
		0.8	0.113E − 04
		0.9	0.133E − 04
		1.0	0.156E − 04
0.0200	0.0002083	0.1	0.396E + 09
		0.2	0.117E + 26
		0.3	0.853E + 42
		0.4	0.416E + 59
		0.5	OVERFLOW
		0.6	OVERFLOW
		0.7	OVERFLOW
		0.8	OVERFLOW
		0.9	OVERFLOW
		1.0	OVERFLOW

First-Order Backward Difference Method

Δx	Δt	t	Maximum Error
0.0200	0.0002	0.1	0.125E − 04
		0.2	0.202E − 04
		0.3	0.265E − 04
		0.4	0.321E − 04
		0.5	0.373E − 04
		0.6	0.426E − 04
		0.7	0.479E − 04
		0.8	0.535E − 04
		0.9	0.596E − 04
		1.0	0.661E − 04
0.0200	0.1000	0.1	0.237E − 02
		0.2	0.430E − 02
		0.3	0.601E − 02
		0.4	0.760E − 02
		0.5	0.913E − 02
		0.6	0.107E − 01
		0.7	0.123E − 01
		0.8	0.140E − 01
		0.9	0.158E − 01
		1.0	0.179E − 01

Fourth-Order Backward Difference Method

Δx	Δt	t	Maximum Error
0.0002	0.0500	0.1	(Specified Exactly)
		0.2	0.208E − 07
		0.3	0.916E − 07
		0.4	0.147E − 06
		0.5	0.193E − 06
		0.6	0.238E − 06
		0.7	0.281E − 06
		0.8	0.324E − 06
		0.9	0.370E − 06
		1.0	0.419E − 06
0.0002	0.1000	0.1	(Specified Exactly)
		0.2	(Specified Exactly)
		0.3	(Specified Exactly)
		0.4	0.654E − 06
		0.5	0.174E − 05
		0.6	0.282E − 05
		0.7	0.370E − 05
		0.8	0.444E − 05
		0.9	0.516E − 05
		1.0	0.594E − 05

the time discretization. When the time step was halved, the error (at $t = 1$) decreased by a factor of 14, approximately confirming the fourth-order accuracy of this time discretization.

If the diffusion equation is nonlinear ($u_t = f(x, t, u, u_x, u_{xx})$), this hardly matters to the explicit methods, since they are still explicit. For implicit methods, however, this means that a system of *nonlinear* equations must be solved each time step. This is generally done using Newton's method (see Section 1.5) or a variation of Newton's method. The Jacobian matrix is still tridiagonal, so several tridiagonal linear systems may be solved each time step (one per Newton iteration).

2.4   Multi-Dimensional Problems

Almost everything said in the previous sections about finite difference solution of one-dimensional diffusion problems generalizes in an obvious way to more than one space dimension. Since it was seen that the lower-order terms in the partial differential equation do not affect stability, and since it will be clear how to treat more general problems, including three-dimensional problems, after studying the following, we will limit our study to the simple two-dimensional diffusion equation

$$u_t = Du_{xx} + Du_{yy} + f(x, y, t),$$

$$u(0, y, t) = g_1(y, t), \qquad 0 \le t,$$

$$u(1, y, t) = g_2(y, t),$$

$$u(x, 0, t) = g_3(x, t) \tag{2.4.1}$$

$$u(x, 1, t) = g_4(x, t),$$

$$u(x, y, 0) = h(x, y), \qquad 0 \le x \le 1, \quad 0 \le y \le 1,$$

where D is assumed to be a positive constant.

A rectangular grid is marked off in the unit square by using the grid lines

$$x_i = i \, \Delta x \qquad \left(\Delta x = \frac{1}{N} \right),$$

$$y_j = j \, \Delta y \qquad \left(\Delta y = \frac{1}{M} \right),$$

and we likewise define

$$t_k = k \, \Delta t.$$

The space grid at different time levels is indicated in Figure 2.4.1. Points where the approximate solution $U(x_i, y_j, t_k)$ is known from the initial or boundary conditions are circled.

The explicit method 2.1.3 for one-dimensional problems generalizes in an obvious way:

$$\frac{U(x_i, y_j, t_{k+1}) - U(x_i, y_j, t_k)}{\Delta t}$$

$$= D \frac{U(x_{i+1}, y_j, t_k) - 2U(x_i, y_j, t_k) + U(x_{i-1}, y_j, t_k)}{\Delta x^2}$$

$$+ D \frac{U(x_i, y_{j+1}, t_k) - 2U(x_i, y_j, t_k) + U(x_i, y_{j-1}, t_k)}{\Delta y^2}$$

$$+ f(x_i, y_j, t_k). \tag{2.4.2}$$

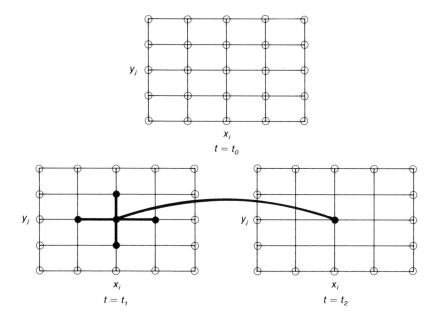

Figure 2.4.1

This can still be explicitly solved for $U(x_i, y_j, t_{k+1})$ in terms of values of U at the previous time level:

$$U(x_i, y_j, t_{k+1}) = [1 - 2D\,\Delta t/\Delta x^2 - 2D\,\Delta t/\Delta y^2]U(x_i, y_j, t_k)$$
$$+ D\,\Delta t/\Delta x^2[U(x_{i+1}, y_j, t_k) + U(x_{i-1}, y_j, t_k)]$$
$$+ D\,\Delta t/\Delta y^2[U(x_i, y_{j+1}, t_k) + U(x_i, y_{j-1}, t_k)]$$
$$+ \Delta t\,f(x_i, y_j, t_k).$$

This formula allows us to calculate U at precisely those values of (x_i, y_j, t_k) where it is needed (uncircled points in Figure 2.4.1); first for $t = t_1$, then t_2, etc. The stencil shows which solution values are related by the explicit formula 2.4.2.

The truncation error for 2.4.2 is found from equations 1.2.5b–c, where it is understood that u is a function of x, y, and t, so that, for example, in formula 1.2.5c, $x = x_i$ and $y = y_j$ are constant throughout. Then

$$T^k_{ij} = \frac{1}{2}\,\Delta t\,u_{tt}(x_i, y_j, \xi_{ijk}) - \frac{D}{12}\,\Delta x^2 u_{xxxx}(\alpha_{ijk}, y_j, t_k) - \frac{D}{12}\,\Delta y^2 u_{yyyy}(x_i, \beta_{ijk}, t_k).$$

Thus the explicit method 2.4.2 is consistent with the equation 2.4.1 and has truncation error $O(\Delta t) + O(\Delta x^2) + O(\Delta y^2)$, assuming the indicated derivatives are bounded.

To determine when this finite difference method is stable, we find the equation satisfied by the error $e \equiv U - u$. Following the procedure used in the previous sections leads to

$$
\begin{aligned}
e(x_i, y_j, t_{k+1}) = {} & [1 - 2D\,\Delta t/\Delta x^2 - 2D\,\Delta t/\Delta y^2]e(x_i, y_j, t_k) \\
& + D\,\Delta t/\Delta x^2[e(x_{i+1}, y_j, t_k) + e(x_{i-1}, y_j, t_k)] \\
& + D\,\Delta t/\Delta y^2[e(x_i, y_{j+1}, t_k) + e(x_i, y_{j-1}, t_k)] \\
& - \Delta t\,T_{ij}^k
\end{aligned}
$$

with

$$
\begin{aligned}
e(x_0, y_j, t_k) &= 0, \\
e(x_N, y_j, t_k) &= 0, \\
e(x_i, y_0, t_k) &= 0, \\
e(x_i, y_M, t_k) &= 0,
\end{aligned}
$$

and

$$
e(x_i, y_j, t_0) = 0.
$$

Now we define $T_{\max}$ and E_k by

$$
T_{\max} = \max_{i,j,k} |T_{ij}^k|,
$$

$$
E_k = \max_{0 \le i \le N, 0 \le j \le M} |e(x_i, y_j, t_k)|.
$$

Then, provided $1 - 2D\,\Delta t/\Delta x^2 - 2D\,\Delta t/\Delta y^2 \ge 0$,

$$
\begin{aligned}
|e(x_i, y_j, t_{k+1})| \le {} & [1 - 2D\,\Delta t/\Delta x^2 - 2D\,\Delta t/\Delta y^2]E_k \\
& + D(\Delta t/\Delta x^2)(E_k + E_k) \\
& + D(\Delta t/\Delta y^2)(E_k + E_k) + \Delta t\,T_{\max} \\
= {} & E_k + \Delta t\,T_{\max}
\end{aligned}
$$

for $1 \le i \le N - 1$ and $1 \le j \le M - 1$. Since $e(x_i, y_j, t_{k+1}) = 0$ when $i = 0$, $i = N, j = 0$, or $j = M$, then

$$
E_{k+1} \le E_k + \Delta t\,T_{\max}.
$$

And since $E_0 = 0$, this implies that

$$E_k \leq k \, \Delta t \, T_{max} = t_k \, T_{max},$$

so the explicit method is stable under the assumption that

$$1 - \frac{2D \, \Delta t}{\Delta x^2} - \frac{2D \, \Delta t}{\Delta y^2} \geq 0$$

or

$$\Delta t \leq \frac{1}{2D(\Delta x^{-2} + \Delta y^{-2})}. \tag{2.4.3}$$

If $\Delta x = \Delta y$, this condition becomes

$$\Delta t \leq \frac{\frac{1}{4} \Delta x^2}{D},$$

which differs from the stability criterion 2.1.7 for the one-dimensional explicit method by only a factor of two.

The implicit methods of Section 2.2 also generalize to the multi-dimensional problem in equally obvious ways. The first-order backward difference method (cf. 2.2.1), for example, is

$$[1 + 2D \, \Delta t / \Delta x^2 + 2D \, \Delta t / \Delta y^2] U(x_i, y_j, t_{k+1})$$

$$- D(\Delta t / \Delta x^2)[U(x_{i+1}, y_j, t_{k+1}) + U(x_{i-1}, y_j, t_{k+1})]$$

$$- D(\Delta t / \Delta y^2)[U(x_i, y_{j+1}, t_{k+1}) + U(x_i, y_{j-1}, t_{k+1})]$$

$$= U(x_i, y_j, t_k) + \Delta t \, f(x_i, y_j, t_{k+1}). \tag{2.4.4}$$

It is easy to show that $E_{k+1} \leq E_k + \Delta t \, T_{max}$ (Problem 6), and thus $E_k \leq t_k \, T_{max}$, so that this implicit method is unconditionally stable as in the one-dimensional case.

The system 2.4.4 of linear equations must be solved at each time step. If $(x_i, \, y_j)$ is not adjacent to the boundary, all the values of U on the left-hand side of 2.4.4 are unknowns, so that this equation involves five unknowns. If $(x_i, \, y_j)$ is adjacent to the boundary, some of the values of U on the left side are known from the boundary conditions. There are a total of $(N - 1)(M - 1)$ equations for the $(N - 1)(M - 1)$ unknowns $U(x_i, y_j, t_{k+1})$, $i = 1, \ldots, N - 1$, $j = 1, \ldots, M - 1$.

Figure 2.4.2 shows the matrix form of the linear system 2.4.4 when the unknowns (and corresponding equations) are ordered in "typewriter" fashion (cf. Figure 0.5.3), and assuming $\Delta x = \Delta y = 0.2$. If the unknowns

$$
\begin{bmatrix}
W & -Z & 0 & 0 & -Z & 0 & 0 & 0 & 0 & 0 & 0 & 0 & 0 & 0 & 0 & 0 \\
-Z & W & -Z & 0 & 0 & -Z & 0 & 0 & 0 & 0 & 0 & 0 & 0 & 0 & 0 & 0 \\
0 & -Z & W & -Z & 0 & 0 & -Z & 0 & 0 & 0 & 0 & 0 & 0 & 0 & 0 & 0 \\
0 & 0 & -Z & W & 0 & 0 & 0 & -Z & 0 & 0 & 0 & 0 & 0 & 0 & 0 & 0 \\
-Z & 0 & 0 & 0 & W & -Z & 0 & 0 & -Z & 0 & 0 & 0 & 0 & 0 & 0 & 0 \\
0 & -Z & 0 & 0 & -Z & W & -Z & 0 & 0 & -Z & 0 & 0 & 0 & 0 & 0 & 0 \\
0 & 0 & -Z & 0 & 0 & -Z & W & -Z & 0 & 0 & -Z & 0 & 0 & 0 & 0 & 0 \\
0 & 0 & 0 & -Z & 0 & 0 & -Z & W & 0 & 0 & 0 & -Z & 0 & 0 & 0 & 0 \\
0 & 0 & 0 & 0 & -Z & 0 & 0 & 0 & W & -Z & 0 & 0 & -Z & 0 & 0 & 0 \\
0 & 0 & 0 & 0 & 0 & -Z & 0 & 0 & -Z & W & -Z & 0 & 0 & -Z & 0 & 0 \\
0 & 0 & 0 & 0 & 0 & 0 & -Z & 0 & 0 & -Z & W & -Z & 0 & 0 & -Z & 0 \\
0 & 0 & 0 & 0 & 0 & 0 & 0 & -Z & 0 & 0 & -Z & W & 0 & 0 & 0 & -Z \\
0 & 0 & 0 & 0 & 0 & 0 & 0 & 0 & -Z & 0 & 0 & 0 & W & -Z & 0 & 0 \\
0 & 0 & 0 & 0 & 0 & 0 & 0 & 0 & 0 & -Z & 0 & 0 & -Z & W & -Z & 0 \\
0 & 0 & 0 & 0 & 0 & 0 & 0 & 0 & 0 & 0 & -Z & 0 & 0 & -Z & W & -Z \\
0 & 0 & 0 & 0 & 0 & 0 & 0 & 0 & 0 & 0 & 0 & -Z & 0 & 0 & -Z & W
\end{bmatrix}
\begin{bmatrix}
U11 \\ U21 \\ U31 \\ U41 \\ U12 \\ U22 \\ U32 \\ U42 \\ U13 \\ U23 \\ U33 \\ U43 \\ U14 \\ U24 \\ U34 \\ U44
\end{bmatrix}
=
\begin{bmatrix}
UK11 + \Delta t*F11 + Z*U01 + Z*U10 \\
UK21 + \Delta t*F21 + Z*U20 \\
UK31 + \Delta t*F31 + Z*U30 \\
UK41 + \Delta t*F41 + Z*U51 + Z*U40 \\
UK12 + \Delta t*F12 + Z*U02 \\
UK22 + \Delta t*F22 \\
UK32 + \Delta t*F32 \\
UK42 + \Delta t*F42 + Z*U52 \\
UK13 + \Delta t*F13 + Z*U03 \\
UK23 + \Delta t*F23 \\
UK33 + \Delta t*F33 \\
UK43 + \Delta t*F43 + Z*U53 \\
UK14 + \Delta t*F14 + Z*U04 + Z*U15 \\
UK24 + \Delta t*F24 + Z*U25 \\
UK34 + \Delta t*F34 + Z*U35 \\
UK44 + \Delta t*F44 + Z*U54 + Z*U45
\end{bmatrix}
$$

where $W = 1 + 4D\,\Delta t/\Delta x^2$ $Uij = U(x_i, y_j, t_{k+1})$

$Z = D\,\Delta t/\Delta x^2$ $UKij = U(x_i, y_j, t_k)$

$Fij = f(x_i, y_j, t_{k+1})$

Figure 2.4.2

Linear System for Two Dimensional Implicit Method

are ordered differently, the rows and columns are permuted, and while the matrix is still "sparse" (has mostly zero entries), it may not have the band structure exhibited in Figure 2.4.2. If N and M are large and equal, this $(N - 1)^2$ by $(N - 1)^2$ matrix will have a "half-bandwidth" of $N - 1$, where the half-bandwidth is defined as $L \equiv \{\max |i - j|$ such that element $\alpha_{ij} \neq 0\}$. Since $W > 4Z$ (W and Z as defined in Figure 2.4.2), this matrix is diagonal dominant, so by Theorem 0.2.1, Gaussian elimination without pivoting may be used to solve the linear system.

It was shown in Section 0.4 that the amount of work required to solve a banded linear system using a band solver (e.g., Figure 0.4.3) is approximately equal to the number of unknowns, $(N - 1)^2$ for this problem, times the square of the half-bandwidth, $L = N - 1$. Thus the work is $O(N^4)$. This is now much greater than the work for one step using the explicit method 2.4.2, which is proportional only to the number of equations, or $O(N^2)$. In higher dimensions, implicitness becomes even more of a handicap. Fortunately, there are other methods for solving the linear system that are more efficient than band elimination (see, for example, Sections 0.5, 4.6, and 4.8). In addition, since the matrix may change little (or not at all, as in our problem) each step, this can further reduce the work, as discussed toward the end of Section 1.5.

The finite difference methods discussed in this section can be easily applied on more general two-dimensional regions only if the boundaries are all parallel to the axes, i.e., only if the region is the union of rectangles. Approximation of boundary conditions along a general, curved, boundary by using finite differences is very cumbersome, though it can be done. It is also awkward, using finite differences, to make the spatial grid nonuniform to accommodate partial differential equations whose solutions vary much more rapidly in certain parts of the region than in others.

Fortunately, there is another class of methods, called "finite element" methods (Chapter 5), that solve partial differential equations in general multi-dimensional regions in a much more natural manner and that also adapt themselves to nonsmooth solutions in a more natural way. We will therefore not waste time detailing the complicated techniques required to use finite differences efficiently in these situations.

2.5 A Diffusion–Reaction Example

Consider the following diffusion–reaction problem, which models the density of di-vacancies (u) and vacancies (v) in a metal, if two vacancies unite to form a di-vacancy with frequency proportional to v^2 and a di-vacancy breaks up to form two vacancies with frequency proportional to u,

$$u_t = u_{xx} + u_{yy} + v^2 - 3u,$$
$$v_t = 4v_{xx} + 4v_{yy} - 2v^2 + 6u.$$

We assume the following boundary conditions on the unit square:

$$\partial u/\partial n = \partial v/\partial n = 0$$

$u = v = 1$ ⬚ $u = v = 1$

$$u = v = 1$$

and the initial conditions

$$u(x, y, 0) = 1,$$
$$v(x, y, 0) = 1.$$

An explicit finite difference discretization is (for $i = 1, \ldots, N - 1$, $j = 1, \ldots, N$)

$$\frac{U(x_i, y_j, t_{k+1}) - U(x_i, y_j, t_k)}{\Delta t}$$
$$= V(x_i, y_j, t_k)^2 - 3U(x_i, y_j, t_k)$$
$$+ \frac{U(x_{i+1}, y_j, t_k) - 2U(x_i, y_j, t_k) + U(x_{i-1}, y_j, t_k)}{\Delta x^2}$$
$$+ \frac{U(x_i, y_{j+1}, t_k) - 2U(x_i, y_j, t_k) + U(x_i, y_{j-1}, t_k)}{\Delta y^2},$$

$$\frac{V(x_i, y_j, t_{k+1}) - V(x_i, y_j, t_k)}{\Delta t}$$
$$= -2V(x_i, y_j, t_k)^2 + 6U(x_i, y_j, t_k)$$
$$+ 4\frac{V(x_{i+1}, y_j, t_k) - 2V(x_i, y_j, t_k) + V(x_{i-1}, y_j, t_k)}{\Delta x^2}$$
$$+ 4\frac{V(x_i, y_{j+1}, t_k) - 2V(x_i, y_j, t_k) + V(x_i, y_{j-1}, t_k)}{\Delta y^2},$$

```
      IMPLICIT DOUBLE PRECISION(A-H,O-Z)
C                              N = NUMBER OF POINTS IN X,Y DIRECTIONS
C                              M = NUMBER OF TIME STEPS
      PARAMETER (N=20, M=3200)
      DIMENSION UK(0:N,0:N+1),UKP1(0:N,0:N+1)
      DIMENSION VK(0:N,0:N+1),VKP1(0:N,0:N+1)
      DX = 1.D0/N
      DY = DX
      DT = 0.5D0/M
C                              BOUNDARY AND INTERIOR VALUES SET TO 1.
C                              (BOUNDARY VALUES WILL NOT CHANGE)
      DO 5 I=0,N
      DO 5 J=0,N+1
        UK(I,J) = 1.0
        VK(I,J) = 1.0
        UKP1(I,J) = 1.0
        VKP1(I,J) = 1.0
    5 CONTINUE
C                              BEGIN MARCHING FORWARD IN TIME
      DO 30 K=1,M
        DO 10 I=1,N-1
        DO 10 J=1,N
          UKP1(I,J) = UK(I,J) + DT*(VK(I,J)**2-3*UK(I,J)
     &    +    (UK(I+1,J)-2*UK(I,J)+UK(I-1,J))/DX**2
     &    +    (UK(I,J+1)-2*UK(I,J)+UK(I,J-1))/DY**2 )
          VKP1(I,J) = VK(I,J) + DT*( - 2*VK(I,J)**2 + 6*UK(I,J)
     &    + 4.*(VK(I+1,J)-2*VK(I,J)+VK(I-1,J))/DX**2
     &    + 4.*(VK(I,J+1)-2*VK(I,J)+VK(I,J-1))/DY**2 )
   10   CONTINUE
C                              HANDLE NORMAL DERIVATIVE BOUNDARY
C                              CONDITION AT Y=1.
        DO 15 I=1,N-1
          UKP1(I,N+1) = UKP1(I,N-1)
          VKP1(I,N+1) = VKP1(I,N-1)
   15   CONTINUE
C                              UPDATE UK,VK
        DO 20 I=0,N
        DO 20 J=0,N+1
          UK(I,J)=UKP1(I,J)
          VK(I,J)=VKP1(I,J)
   20   CONTINUE
        TKP1 = K*DT
C                              PRINT SOLUTION AT (0.5,0.5)
        PRINT 25, TKP1,UKP1(N/2,N/2),VKP1(N/2,N/2)
   25   FORMAT (3E15.5)
   30 CONTINUE
      STOP
      END
```

Figure 2.5.1
FORTRAN77 Program to Solve Diffusion–Reaction Problem

with

$$U(x_0, y_j, t_k) = 1, \qquad\qquad V(x_0, y_j, t_k) \ = 1,$$
$$U(x_N, y_j, t_k) = 1, \qquad\qquad V(x_N, y_j, t_k) \ = 1,$$
$$U(x_i, y_0, t_k) = 1, \qquad\qquad V(x_i, y_0, t_k) \ = 1,$$
$$U(x_i, y_{N+1}, t_k) = U(x_i, y_{N-1}, t_k), \qquad V(x_i, y_{N+1}, t_k) = V(x_i, y_{N-1}, t_k),$$

and

$$U(x_i, y_j, t_0) = 1, \qquad\qquad V(x_i, y_j, t_0) \ = 1,$$

where $x_i = i\,\Delta x$, $y_j = j\,\Delta y$, $t_k = k\,\Delta t$, and $\Delta x = \Delta y = 1/N$.

The implementation of this explicit method is simple, and a FORTRAN77 program, with $\Delta x = \Delta y = 0.05$, $\Delta t = 1.5625\mathrm{E} - 4$, is shown in Figure 2.5.1.

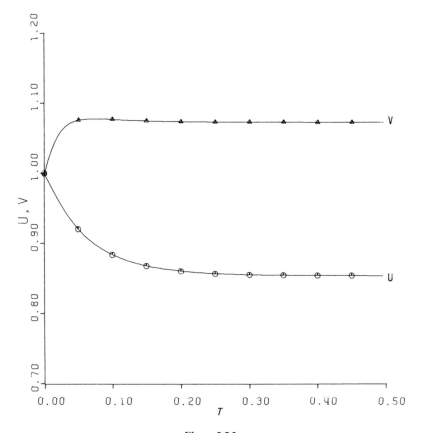

Figure 2.5.2
Time Behavior of U and V at Midpoint

The value chosen for Δt is the largest for which the stability criterion 2.4.3 is satisfied for both $D = 1$ and $D = 4$. An attempt with Δt slightly larger, $\Delta t = 1.6667E - 4$, led to instability and overflow. The values of u and v at the midpoint of the square, as functions of time, are shown in Figure 2.5.2. By $t = 0.5$ a steady-state condition appears to have been reached. The values of u and v at $x = 0.5$, $y = 0.5$, $t = 0.5$ are calculated as $u = 0.8537$, $v = 1.0732$, in good agreement with the values $u = 0.8535$, $v = 1.0733$ calculated by PDE/ PROTRAN, a finite element program first discussed in Section 5.6.

   Although the use of an explicit method forced us to take a large number of time steps, using an implicit method would have complicated the programming tremendously. The fact that this problem is nonlinear, involves two partial differential equations and two space dimensions, and has mixed boundary conditions hardly affects the programming complexity when an explicit method is used, but all of these factors complicate things for an implicit method. Unless extremely high accuracy is required, making the computations very expensive, these programming considerations usually make the explicit methods more attractive than the (possibly more efficient) implicit methods.

2.6   Problems

1. Verify that if u_1 and u_2 are two solutions of the diffusion problem (cf. 2.0.3),

$$u_t = \mathbf{V}^T(D(x, y, z, t) \, \mathbf{Vu}) + f(x, y, z, t) \qquad \text{in R,}$$

$$u = g_0(x, y, z, t) \qquad \text{on part of the boundary,}$$

$$\partial u / \partial u = g_1(x, y, z, t) \qquad \text{on the other part,}$$

$$u = h(x, y, z) \qquad \text{at } t = 0,$$

where $D > 0$, then their difference, $e \equiv u_1 - u_2$, satisfies

$$(e^2)_t = 2 \, \mathbf{V}^T(eD \, \mathbf{Ve}) - 2D|\mathbf{Ve}|^2.$$

Using this result, show that there is at most one (smooth) solution to this diffusion problem.
Hint: Show that the integral over R of e^2 is nonincreasing and is zero when $t = 0$.

2. Show that u given by 2.0.6 satisfies $u_t = D \nabla^2 u$ and that

$$\int_{-\infty}^{\infty} \int_{-\infty}^{\infty} \int_{-\infty}^{\infty} g(x - \alpha, y - \beta, z - \Gamma, t) \, d\alpha \, d\beta \, d\Gamma = 1.$$

Hint: Use the fact that $\displaystyle\int_{-\infty}^{\infty} \exp(-x^2) \, dx = \sqrt{\pi}$.

3. Assuming that $D > 0$ is constant and $a \equiv v \equiv 0$, if the ordinary differential equation system 2.1.8 is written in the form $\mathbf{U}' = B\mathbf{U} + \mathbf{b}$, show that the eigenvalues of B are $\lambda_m = -4D/\Delta x^2 \sin^2(\frac{1}{2}m\pi\Delta x)$, for $m = 1, \ldots,$ $N - 1$, so that $\lambda_1 \approx -D\pi^2$ and $\lambda_{N-1} \approx -4D/\Delta x^2$. Then argue that even if a and v are nonzero, 2.1.8 is stiff and becomes more and more stiff as $\Delta x \to 0$.
Hint: $\mathbf{U_m} = (\sin(m\pi x_1), \ldots, \sin(m\pi x_{N-1}))$ is an eigenvector of B, for $m = 1, \ldots, N - 1$. The homogeneous portion ($f = g_0 = g_1 = 0$) of the right-hand side of 2.1.8 represents $B\mathbf{U}$, so, to verify that $\mathbf{U_m}$ is an eigenvector, simply replace U_i in the homogeneous version of 2.1.8 by the ith component of $\mathbf{U_m}$ and verify that the ith component of $B\mathbf{U_m}$ is a constant multiple of the ith component of $\mathbf{U_m}$. The fact that $\sin(m\pi x_0) = \sin(m\pi x_N) = 0$ for any m makes it unnecessary to treat the cases $i = 1$ and $i = N - 1$ separately. Then use Problem 9 of Chapter 1.

4. Use the Crank–Nicolson method to solve the partial differential equation 2.3.1, that is, use the second-order implicit Adams–Moulton method (Table 1.4.1, $m = 1$) to solve the ordinary differential equation system 2.3.2. You may use the subroutine in Figure 0.4.4 to solve the tridiagonal linear system each step. Use $\Delta x = 0.02$, $\Delta t = 0.1$, and compare your accuracy with the results obtained by using the first-order backward difference method with the same values of Δx and Δt, as shown in Table 2.3.1.

5. Use a first-order implicit method similar to 2.2.1 to solve the nonlinear problem

$$u_t = u_{xx} + u^2 - u - x^2 e^{-2t},$$

$$u(0, t) = 0, \qquad 0 \le t \le 1,$$

$$u(1, t) = e^{-t},$$

$$u(x, 0) = x, \qquad 0 \le x \le 1,$$

which has exact solution $u = xe^{-t}$. At each time step, you will have to solve a system of *nonlinear* equations of the form $\mathbf{f}(\mathbf{U}) = \mathbf{0}$ for the vector of unknowns $\mathbf{U} = (U(x_1, t_{k+1}), \ldots, U(x_{N-1}, t_{k+1}))$. Use Newton's iteration:

$$\mathbf{U^{n+1}} = \mathbf{U^n} - \mathbf{\Delta^n},$$

where

$$J^n\mathbf{\Delta^n} = \mathbf{f^n},$$

to solve this nonlinear system. Here J represents the Jacobian matrix of $\mathbf{f}$, and the superscript n on J and $\mathbf{f}$ means that their components are evaluated at $\mathbf{U^n}$. The Jacobian matrix will be tridiagonal, so the FORTRAN subroutine in Figure 0.4.6 may be used to solve for $\mathbf{\Delta^n}$ each iteration. If the solution values at t_k are used to start the iteration, two or three Newton iterations per time step are sufficient.

Use $\Delta x = \Delta t = 0.05$ and compare your approximate solution with the exact solution at $t = 1$.

6. Show that $E_{k+1} \leq E_k + \Delta t\, T_{\max}$, where

$$E_k \equiv \max_{0 \leq i \leq N, 0 \leq j \leq M} |U(x_i, y_j, t_k) - u(x_i, y_j, t_k)|$$

and where u is the solution to the partial differential equation 2.4.1, U is the solution to 2.4.4, and $T_{\max}$ is a bound on the truncation error. This shows that the implicit method applied to 2.4.1 is unconditionally stable.

7. Consider the partial differential equation

$$u_t = u_{xx} + u_{yy} + u + 4e^t \qquad \text{in the unit circle,}$$

with

$$u = 0 \qquad\qquad \text{on the boundary,}$$
$$u = 1 - x^2 - y^2 \qquad \text{at } t = 0.$$

In polar coordinates ($x = r \cos \theta$, $y = r \sin \theta$), this partial differential equation takes the form

$$u_t = u_{rr} + \frac{u_r}{r} + \frac{u_{\theta\theta}}{r^2} + u + 4e^t.$$

In the new variables, the solution domain is rectangular, $0 \leq \theta \leq 2\pi$, $0 \leq r \leq 1$, and this makes the polar coordinate form more tractable.

Discretize this form of the partial differential equation by using central differences to approximate u_{rr}, u_r, and $u_{\theta\theta}$ and a forward difference to approximate u_t, giving an explicit finite difference method.

At $\theta = 0$ and $\theta = 2\pi$ there are no boundary conditions, but this poses no problem, provided we recall that $U(\theta_i, r_j) = U(\theta_{N+i}, r_j)$ for any integer i, where $\theta_i = i\,\Delta\theta$, $r_j = j\,\Delta r$, $\Delta\theta = 2\pi/N$, $\Delta r = 1/M$. At $r = 1$ the values of U are known (zero), but at $r = 0$ we appear to have a problem, since not only do we have no boundary conditions there, but the partial differential equation in polar form has terms of the form $1/r$ in it. The easiest way to avoid both problems is to use the discretized Cartesian form at $r = 0$ only, which results in an equation for U at $r = 0$ in terms of its values on the previous time level at $r = r_1$, $\theta = \{0, 0.5\pi, \pi, 1.5\pi\}$. Thus N must be a multiple of 4, so that these points will belong to the (θ_i, r_j) grid.

With $\Delta\theta = 2\pi/20$, $\Delta r = 0.1$, $\Delta t = 0.002$, solve the partial differential equation from $t = 0$ to $t = 1$, and compare your results with the true solution $u = e^t(1 - r^2)$. How large can you make Δt before the explicit method becomes unstable?

8. Use the Fourier stability method outlined in Appendix 1 to determine the conditions under which the following approximations to $u_t = u_{xx}$ are stable.

a. $$\frac{U(x_i, t_{k+1}) - U(x_i, t_{k-1})}{2\,\Delta t} = \frac{U(x_{i+1}, t_k) - 2U(x_i, t_k) + U(x_{i-1}, t_k)}{\Delta x^2}.$$

b. $$\frac{U(x_i, t_{k+1}) - U(x_i, t_{k-1})}{2\,\Delta t}$$
$$= \frac{U(x_{i+1}, t_k) - U(x_i, t_{k+1}) - U(x_i, t_{k-1}) + U(x_{i-1}, t_k)}{\Delta x^2}.$$

Note that although formula (b) is unconditionally stable, it is consistent with $u_t = u_{xx}$ only if $\Delta t/\Delta x \to 0$, so this explicit method still must be used with small values of Δt.

3

The Initial Value
Transport and Wave Problems

3.0 Introduction

Despite the fact that the transport equation is first-order and the wave problem is second-order, these two problems are grouped together in this chapter because they have many properties in common, and both are classified as "hyperbolic" problems. The most striking characteristic common to both problems is that, unlike the "parabolic" and "elliptic" problems discussed in Chapters 2 and 4, respectively, solutions to these hyperbolic problems often are very ill-behaved or even discontinuous. This naturally makes their approximation difficult.

The "transport" problem is simply a diffusion–convection problem 2.0.3 where diffusion is negligibly slow compared to convection:

$$u_t = -\nabla^T(u\mathbf{v}) + f \qquad (3.0.1)$$

or

$$u_t = -(uv_1)_x - (uv_2)_y - (uv_3)_z + f,$$

where $\mathbf{v} = (v_1, v_2, v_3)$ is the convection velocity and f is the source term. The variables v_1, v_2, v_3 and the function f may depend on $x, y, z, t,$ and even u.

It is physically clear why the transport equation solutions may be badly behaved. It is the smoothing, or averaging, effect of diffusion that ensures that solutions to the diffusion problem 2.0.3 are smooth (away from the boundary and for $t > 0$), even if the initial or boundary conditions are not. When diffusion is not present, however, the convection current or wind simply carries the initial and boundary conditions along with it, propagating any rough behavior into the interior of the region. From the physical nature of the problem, it is also clear that boundary values for u ought to be prescribed only on that portion of the boundary where the flux is inward, i.e., where $\mathbf{v}^T\mathbf{n}$ is negative ($\mathbf{n}$ = unit outward normal vector). In other words, the concentration ought to be specified upwind from where the concentration is to be calculated, not downwind. The values of u also must be specified throughout the region of interest at $t = 0$. (See Problem 1.)

It is instructive to look at the simplest version of 3.0.1—in one space dimension, with $f = 0$ and $v > 0$ constant:

$$u_t = -vu_x,$$

$$u(0, t) = g(t), \qquad 0 \le t,$$

$$u(x, 0) = h(x), \qquad 0 \le x \le 1.$$

We can think of $u(x, t)$ as representing the concentration of some solute in water flowing through a section of pipe from left to right with velocity v. Thus the boundary condition is given at the inlet end, $x = 0$. Along any line given parametrically by $x = x_0 + vs$, $t = s$ (see Figure 3.0.1), we have

$$\frac{du}{ds} = u_x \frac{dx}{ds} + u_t \frac{dt}{ds} = u_x v + u_t = 0.$$

Therefore, u is constant along these lines, called "characteristics."

This is reasonable physically, as this means that the water simply carries the solute along with it at velocity v, with no diffusion. If we want to know the concentration at (x, t), we can simply reverse the motion of the water and follow it back until it reaches either a point on the initial line $t = 0$ or the left boundary $x = 0$ where the concentration is known. It is clear that specifying the concentration on the right end of the pipe, $x = 1$, where the flux is outward, makes no sense and would either contradict, or be redundant with, the initial conditions.

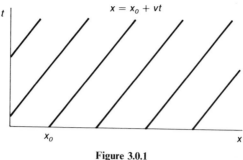

Figure 3.0.1

Now let us consider the vibrating string problem, which is modeled by the one-dimensional wave equation. Suppose that a taut string is suspended by its extremes at the points $(0, g_0)$ and $(1, g_1)$ (Figure 3.0.2) and that a vertical force of $f(x, t)$ per unit length is applied to it at $(x, u(x, t))$, where $u(x, t)$ represents the vertical position of the string.

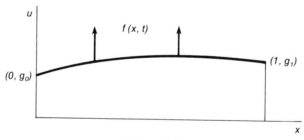

Figure 3.0.2

We will assume that u_x is small for all x and t. This assures that the string is never stretched much from its unloaded length, and therefore the string tension can be taken to be a constant T. Figure 3.0.3 shows the forces acting on a small segment of the string between x and $x + \Delta x$.

If the angle between the string tangent and the horizontal is called $\alpha(x, t)$, then setting the sum of the vertical components of the forces acting on the string segment equal to the product of the mass and vertical acceleration of that segment gives

$$\rho \, \Delta x \, u_{tt} = T \sin \alpha(x + \Delta x, t) - T \sin \alpha(x, t) + \Delta x \, f(x, t),$$

where ρ is the linear density (mass per unit length) of the string.

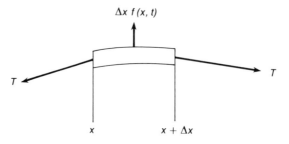

Figure 3.0.3

Since u_x is assumed to be small, we can take

$$\sin \alpha(x, t) \approx \tan \alpha(x, t) = u_x(x, t),$$

and thus

$$\rho u_{tt} = T \frac{u_x(x + \Delta x, t) - u_x(x, t)}{\Delta x} + f(x, t).$$

Finally, taking the limit as $\Delta x \to 0$ gives the "wave equation"

$$\rho u_{tt} = T u_{xx} + f. \tag{3.0.2}$$

There may also be a frictional force acting on this string segment that is proportional to the velocity and in an opposing direction. This introduces a term $-au_t$ into the right-hand side, and we have the "damped" wave equation:

$$\rho u_{tt} = -au_t + T u_{xx} + f. \tag{3.0.3}$$

The multi-dimensional version of this equation is

$$\rho u_{tt} = -au_t + T \nabla^2 u + f. \tag{3.0.4}$$

In two dimensions, it models the displacement $u(x, y, t)$ of an elastic membrane.

The boundary conditions for 3.0.2 or 3.0.3 may be $u(0, t) = g_0, u(1, t) = g_1$, where g_0 and g_1 may possibly be functions of t. For 3.0.4, u may be specified around the entire boundary, but other boundary conditions are possible. Since the wave equation is second-order in time, both the initial position and initial velocity of each portion of the string (or membrane) must be specified, so the initial conditions are (see Problem 1)

$$u = h \qquad \text{at } t = 0,$$

$$u_t = q \qquad \text{at } t = 0.$$

It is instructive to examine the simplest version of the wave equation 3.0.2, on an infinite interval, with $f = 0$:

$$\rho u_{tt} = T u_{xx},$$
$$u(x, 0) = h(x), \qquad -\infty < x < \infty,$$
$$u_t(x, 0) = q(x).$$

This problem has a well-known solution, called the d'Alembert solution (see Problem 2):

$$u(x, t) = \tfrac{1}{2}h(x + ct) + \tfrac{1}{2}h(x - ct) + \frac{1}{2c}\int_{x-ct}^{x+ct} q(s)\, ds, \qquad (3.0.5)$$

where $c \equiv \sqrt{(T/\rho)}$.

Actually, the d'Alembert solution is valid at (x, t) even if the interval is finite, until t is large enough that the interval of dependence $(x - ct, x + ct)$ includes a boundary point.

From 3.0.5 it can be seen that the solution at a given point (x, t) depends only on the initial values between $(x - ct, 0)$ and $(x + ct, 0)$. The values of h and q at points further than ct from x have not had time to affect the solution at x by the time t. Thus c is called the wave velocity, since information travels a distance of ct in time t.

It can also be seen from 3.0.5 that, as claimed earlier, if the initial conditions h and/or q are badly behaved, the solution may be also.

The equations that model displacements in an elastic body, such as a metal object, can be similarly derived using Newton's second law. If we focus on an arbitrary small volume E of this body and set the mass times the acceleration equal to the sum of the internal and external forces acting on E, we get:

$$\iiint_E \rho u_{tt}\, dx\, dy\, dz = \iint_{\partial E} (\sigma_{xx} n_x + \sigma_{xy} n_y + \sigma_{xz} n_z)\, dA + \iiint_E f_1\, dx\, dy\, dz,$$

$$\iiint_E \rho v_{tt}\, dx\, dy\, dz = \iint_{\partial E} (\sigma_{yx} n_x + \sigma_{yy} n_y + \sigma_{yz} n_z)\, dA + \iiint_E f_2\, dx\, dy\, dz,$$

$$\iiint_E \rho w_{tt}\, dx\, dy\, dz = \iint_{\partial E} (\sigma_{zx} n_x + \sigma_{zy} n_y + \sigma_{zz} n_z)\, dA + \iiint_E f_3\, dx\, dy\, dz.$$

$$(3.0.6)$$

Here $\rho(x, y, z)$ is the density, and $(u(x, y, z, t), v(x, y, z, t), w(x, y, z, t))$ is the displacement vector. That is, the body element that is at (x, y, z) when the elastic body is unloaded and in equilibrium is displaced to $(x + u, y + v, z + w)$ at time t. Thus (u_{tt}, v_{tt}, w_{tt}) is the acceleration vector. The stresses $\sigma_{xx} \ldots$ are the elements of a stress tensor (matrix), defined so that

$$\begin{bmatrix} \sigma_{xx} & \sigma_{xy} & \sigma_{xz} \\ \sigma_{yx} & \sigma_{yy} & \sigma_{yz} \\ \sigma_{zx} & \sigma_{zy} & \sigma_{zz} \end{bmatrix} \begin{bmatrix} n_x \\ n_y \\ n_z \end{bmatrix}$$

gives the force per unit area that the rest of the body exerts on a boundary element of E that has unit outward normal (n_x, n_y, n_z). Thus σ_{pq} ($p, q = x, y$ or z) can be interpreted as the component in the p-axis direction of the internal force experienced by a boundary element of unit area that is perpendicular to the q-axis. The vector $(f_1(x, y, z, t), f_2(x, y, z, t), f_3(x, y, z, t))$ represents the force per unit volume attributable to external sources, such as gravity.

After the divergence theorem is applied to the boundary integrals in 3.0.6, we get

$$\iiint_E \rho u_{tt} \, dx \, dy \, dz = \iiint_E (\sigma_{xx})_x + (\sigma_{xy})_y + (\sigma_{xz})_z + f_1 \, dx \, dy \, dz,$$

$$\iiint_E \rho v_{tt} \, dx \, dy \, dz = \iiint_E (\sigma_{yx})_x + (\sigma_{yy})_y + (\sigma_{yz})_z + f_2 \, dx \, dy \, dz,$$

$$\iiint_E \rho w_{tt} \, dx \, dy \, dz = \iiint_E (\sigma_{zx})_x + (\sigma_{zy})_y + (\sigma_{zz})_z + f_3 \, dx \, dy \, dz.$$

Since E is an arbitrary volume, it can be made so small that the integrands are almost constant, so that at any point

$$\rho u_{tt} = (\sigma_{xx})_x + (\sigma_{xy})_y + (\sigma_{xz})_z + f_1,$$
$$\rho v_{tt} = (\sigma_{yx})_x + (\sigma_{yy})_y + (\sigma_{yz})_z + f_2, \qquad (3.0.7)$$
$$\rho w_{tt} = (\sigma_{zx})_x + (\sigma_{zy})_y + (\sigma_{zz})_z + f_3.$$

For a linear, isotropic (direction independent), elastic body, it has been determined experimentally that

$$\sigma_{xx} = \frac{E[(1 - \mu)u_x + \mu v_y + \mu w_z]}{(1 + \mu)(1 - 2\mu)},$$

$$\sigma_{yy} = \frac{E[\mu u_x + (1 - \mu)v_y + \mu w_z]}{(1 + \mu)(1 - 2\mu)},$$

$$\sigma_{zz} = \frac{E[\mu u_x + \mu v_y + (1 - \mu)w_z]}{(1 + \mu)(1 - 2\mu)},$$

$$\sigma_{xy} = \sigma_{yx} = \frac{E[u_y + v_x]}{2(1 + \mu)},$$

$$\sigma_{xz} = \sigma_{zx} = \frac{E[u_z + w_x]}{2(1 + \mu)},$$

$$\sigma_{yz} = \sigma_{zy} = \frac{E[v_z + w_y]}{2(1 + \mu)},$$

where E and μ are material properties called the elastic modulus and the Poisson ratio, respectively.

Therefore, for such a body, 3.0.7 represents a system of second-order partial differential equations, whose properties are qualitatively similar to those of the wave equation 3.0.2.

On the boundary, normally either the displacement vector (u, v, w) is specified, or else an applied external boundary force (per unit area) vector $(g_1(x, y, z, t), g_2(x, y, z, t), g_3(x, y, z, t))$ is given. The boundary conditions that model the latter situation are obtained by balancing the internal and external forces on the boundary:

$$\begin{bmatrix} \sigma_{xx} & \sigma_{xy} & \sigma_{xz} \\ \sigma_{yx} & \sigma_{yy} & \sigma_{yz} \\ \sigma_{zx} & \sigma_{zy} & \sigma_{zz} \end{bmatrix} \begin{bmatrix} n_x \\ n_y \\ n_z \end{bmatrix} = \begin{bmatrix} g_1 \\ g_2 \\ g_3 \end{bmatrix}.$$

At $t = 0$, the initial displacements (u, v, w) and the initial displacement velocities (u_t, v_t, w_t) must be specified throughout the body.

3.1 Explicit Methods for the Transport Problem

Let us first consider the one-dimensional version of the transport equation 3.0.1. If the problem is linear, then it can be put into the general form

$$u_t = -v(x, t)u_x + a(x, t)u + f(x, t),$$

$$u(0, t) = g(t), \qquad 0 \le t \le t_f, \qquad\qquad (3.1.1)$$

$$u(x, 0) = h(x), \qquad 0 \le x \le 1.$$

We assume that $v(x, t)$ is positive, so that the flux is inward on the left end of the interval $(0, 1)$, and thus it is appropriate to assign boundary values at $x = 0$.

For later use, we define

$$V_{max} = \max |v(x, t)|, \qquad A_{max} = \max |a(x, t)|,$$

where the maxima are taken over $0 \le x \le 1, 0 \le t \le t_f$.

It seems reasonable (at first!) to use the same central difference formulas as used for the diffusion–convection problem 2.1.1, that is, to use formulas 2.1.3 without the diffusion term ($D = 0$):

$$\frac{U(x_i, t_{k+1}) - U(x_i, t_k)}{\Delta t} = -v(x_i, t_k) \frac{U(x_{i+1}, t_k) - U(x_{i-1}, t_k)}{2 \Delta x}$$

$$+ a(x_i, t_k)U(x_i, t_k) + f(x_i, t_k), \qquad (3.1.2)$$

$$U(x_0, t_k) = g(t_k),$$

$$U(x_i, t_0) = h(x_i).$$

This explicit formula can be solved for $U(x_i, t_{k+1})$ in terms of values of U at the previous time level t_k (formula 2.1.4 with $D = 0$). But when we look at Figure 3.1.1, which shows the stencil corresponding to the central difference formula 3.1.2 and marks those values of U that are known from the boundary or initial conditions with circles, we see that there is a problem. As for the diffusion problem, the difference formulas can only calculate the values of U at the interior points, so that $U(x_N, t_1)$, for example, (and then $U(x_{N-1}, t_2)$, $U(x_N, t_2)$) cannot be calculated. For the diffusion–convection problem, these were supplied by the second boundary condition, which does not exist now.

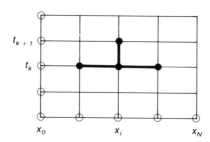

Figure 3.1.1
Central Difference Stencil

In fact, at level t_k, with the central difference formula, only the values of U at $x_0, x_1, \ldots, x_{N-k}$ can be calculated, and so for a fixed Δx, the smaller Δt is, the smaller the interval on which the solution at a given time value can be calculated. In any case, the central difference formula 3.1.2 can be shown to be unconditionally unstable (see Appendix 1), so it will not be considered further.

Because of these problems, we prefer to use "upwind," rather than central differences, to approximate the u_x term in 3.1.1. That is,

$$\frac{U(x_i, t_{k+1}) - U(x_i, t_k)}{\Delta t} = -v(x_i, t_k)\frac{U(x_i, t_k) - U(x_{i-1}, t_k)}{\Delta x}$$

$$+ a(x_i, t_k)U(x_i, t_k) + f(x_i, t_k), \qquad (3.1.3)$$

$$U(x_0, t_k) = g(t_k),$$

$$U(x_i, t_0) = h(x_i).$$

This formula is still explicit, but the value of $U(x_i, t_{k+1})$ is now calculated in terms of $U(x_i, t_k)$ and $U(x_{i-1}, t_k)$ only (see the stencil for 3.1.3 in Figure 3.1.2). Now there is no problem calculating all the unknown values of U, first at $t = t_1$, then at $t = t_2$, etc., including the values at $x = x_N$.

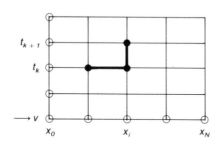

Figure 3.1.2
Upwind Difference Stencil

To analyze the stability of 3.1.3, we follow the procedure established in Chapter 2 and note that the exact solution of 3.1.1, $u(x, t)$, satisfies a difference equation like 3.1.3 but with an extra term, the truncation error, which is now (cf. 1.2.5c)

$$T_i^k = \tfrac{1}{2}\Delta t \, u_{tt}(x_i, \xi_{ik}) - \tfrac{1}{2}\Delta x \, v(x_i, t_k)u_{xx}(\alpha_{ik}, t_k) = O(\Delta t) + O(\Delta x).$$

Subtracting the equation satisfied by u from that satisfied by U yields a difference equation for the error $e \equiv U - u$.

$$\frac{e(x_i, t_{k+1}) - e(x_i, t_k)}{\Delta t} = -v(x_i, t_k) \frac{e(x_i, t_k) - e(x_{i-1}, t_k)}{\Delta x}$$

$$+ a(x_i, t_k)e(x_i, t_k) - T_i^k \qquad (i = 1, \ldots, N),$$

$$e(x_0, t_k) = 0,$$

$$e(x_i, t_0) = 0,$$

or

$$e(x_i, t_{k+1}) = [1 - (\Delta t/\Delta x)v(x_i, t_k) + \Delta t\, a(x_i, t_k)]e(x_i, t_k)$$

$$+ (\Delta t/\Delta x)v(x_i, t_k)e(x_{i-1}, t_k) - \Delta t\, T_i^k. \qquad (3.1.4)$$

We define, as before, $T_{\max} \equiv \max |T_i^k|$ and

$$E_k \equiv \max_{0 \le i \le N} |e(x_i, t_k)|.$$

Then,

$$|e(x_i, t_{k+1})| \le |1 - (\Delta t/\Delta x)v(x_i, t_k)| E_k + (\Delta t/\Delta x)|v(x_i, t_k)| E_k$$

$$+ \Delta t\, A_{\max}\, E_k + \Delta t\, T_{\max}.$$

If $1 - (\Delta t/\Delta x)V_{\max} \ge 0$, then (since $v \ge 0$) the first two terms under the absolute value signs on the right are nonnegative, so that the absolute value signs may be removed, and

$$|e(x_i, t_{k+1})| \le (1 + \Delta t\, A_{\max})E_k + \Delta t\, T_{\max} \qquad (i = 1, \ldots, N).$$

Since $e(x_0, t_{k+1}) = 0$,

$$E_{k+1} \le (1 + \Delta t\, A_{\max})E_k + \Delta t\, T_{\max}.$$

Applying Theorem 1.2.1 to this inequality, since $E_0 = 0$, gives

$$E_k \le \exp(A_{\max} t_k)t_k T_{\max},$$

so the upwind difference formula 3.1.3 is stable, under the assumption that $1 - (\Delta t/\Delta x)V_{\max} \ge 0$, that is, that

$$\Delta t \le \frac{\Delta x}{V_{\max}}. \qquad (3.1.5)$$

This stability restriction makes sense when the "domain of dependence" of this explicit method is considered, at least when v is constant. If Δt and Δx go to zero with a fixed ratio $\Delta t/\Delta x = r$, the value of the approximate solution at a fixed point (x, t) will depend only on the initial values between $x - t/r$ and x (see Figure 3.1.3). Thus as Δt and Δx go to zero, with constant ratio, all the approximate solutions, and therefore their limit, depend only on the values of $h(x)$ within the base of the triangle drawn.

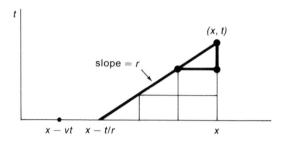

Figure 3.1.3

The true solution at (x, t) depends on the value of $h(x)$ only at the point $x - vt$. This can be seen by noting that along the lines given parametrically by $x = x_0 + vs$, $t = s$, the partial differential equation 3.1.1 (with $v = $ constant) reduces to the ordinary differential equation

$$\frac{du}{ds} = a(x_0 + vs, s)u + f(x_0 + vs, s). \tag{3.1.6}$$

Thus the solution at any point (x, t) can be found by solving the first-order ordinary differential equation 3.1.6 with initial value $u(0) = u(x_0, 0) = h(x_0)$. But if $(x_0 + vs, s)$ passes through (x, t), we find, by equating these points, that $x_0 = x - vt$, so the solution at (x, t) depends only on $h(x - vt)$.

If $x - vt$ lies outside the interval $(x - t/r, x)$, i.e., if $v > 1/r = \Delta x/\Delta t$, or $\Delta t > \Delta x/v$, then the approximate solutions cannot possibly be converging to the true solution, in general, because we can change the initial value at $x - vt$, changing the true solution but not the limit of the approximate solutions. Thus for stability, $\Delta t \leq \Delta x/v$, in agreement with 3.1.5, when v is constant.

(If, on the other hand, $v < 1/r$, the student may draw the incorrect conclusion that the method is still unstable, because the initial values may be changed at points inside the domain of dependence of the approximate solution, but not at $x - vt$, thereby changing the approximate solutions and not the true solution. The error in this reasoning is that changing the approximate solutions does not necessarily change their limit.)

If, instead, u_x is approximated by the "downwind" difference scheme (see Figure 3.1.4),

$$u_x \approx \frac{U(x_{i+1}, t_k) - U(x_i, t_k)}{\Delta x},$$

the approximate solutions at (x, t) depend only on the initial conditions between x and $x + t/r$, that is, on what happened downwind at earlier times. Clearly this is an unconditionally *unstable* scheme, because no matter how small Δt is, the point $(x - vt, 0)$ on whose initial value the true solution depends is never in the domain of dependence of the approximate solutions. Of course, it makes no sense to calculate the solution at (x, t) based on what happened earlier downwind, since only the upwind concentration (or temperature) is going to affect the concentration at later times. This scheme is useless anyway, since it tells us how to calculate the solution values at x_0, where they are already known, and not at x_N, where they are needed!

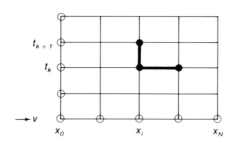

Figure 3.1.4
Downwind Difference Stencil

It is interesting to notice that using an upwind difference approximation to u_x is equivalent to adding an "artificial diffusion" term $\frac{1}{2}|v|\Delta x \, u_{xx}$ to the transport problem 3.1.1 and to using the central difference scheme 2.1.3 to approximate this new diffusion–convection problem (Problem 3). In other words, upwinding has the same smoothing effect as adding a small amount of diffusion to the convection model. Since the diffusion coefficient goes to zero with Δx, the limit of the approximate solutions does not change.

The stability condition $\Delta t \leq \Delta x / V_{max}$ for the convection problem using 3.1.3 is a much less severe restriction on Δt than the condition $\Delta t \leq \frac{1}{2} \Delta x^2 / D_{max}$ (formula 2.1.7) for the diffusion–convection problem using the explicit method 2.1.3. (Equations 3.1.3 represent the application of Euler's method to an ordinary differential equation system that is stiff but not as stiff as the diffusion system 2.1.8.) Removing this restriction by introducing implicit methods is therefore less critical for the transport equation, so implicit methods will not be pursued here. As shown in Appendix 1, however, unconditionally stable implicit methods do exist. In fact, the upwind "implicit" method considered there requires only the solution of a bidiagonal (triangular) matrix each time step, so that this method is—for practical purposes—explicit!

As an example, consider the nonlinear transport problem

$$u_t = 2u_x - u^2 + \exp[4t + 2x],$$

$$u(1, t) = \exp[2t + 1], \qquad 0 \leq t \leq 1, \tag{3.1.7}$$

$$u(x, 0) = \exp[x], \qquad 0 \leq x \leq 1,$$

which has the exact solution $u(x, t) = \exp[2t + x]$. Notice that $v = -2$, so that the boundary condition must be given on the right end of the interval, where the current is inward. We solve 3.1.7 by using the upwind difference formula ("upwind" now means biased toward $x = 1$)

$$\frac{U(x_i, t_{k+1}) - U(x_i, t_k)}{\Delta t} = 2 \frac{U(x_{i+1}, t_k) - U(x_i, t_k)}{\Delta x}$$

$$- U(x_i, t_k)^2 + \exp[4t_k + 2x_i].$$

Table 3.1.1 shows that when Δt is less than $\Delta x / V_{max} = \frac{1}{2} \Delta x$, the formula is stable, but when it exceeds this limit, it becomes useless.

Now let us consider the *system* of m first-order hyperbolic equations

$$\mathbf{u}_t = -V\mathbf{u}_x + \mathbf{f}(x, t),$$

$$\mathbf{u}(0, t) = \mathbf{g}(t), \qquad 0 \leq t \leq t_f,$$

$$\mathbf{u}(x, 0) = \mathbf{h}(x), \qquad 0 \leq x \leq 1. \tag{3.1.8}$$

This constant coefficient problem is, by definition, hyperbolic if $V = P^{-1}DP$, where D is a real diagonal matrix [Mitchell and Griffiths, 1980, Section 4.6]. The system then reduces to a set of uncoupled equations $(z_i)_t = -\lambda_i(z_i)_x + q_i(x, t)$, where the z_i and q_i are the components of $\mathbf{z} = P\mathbf{u}$

Table 3.1.1
Errors for Explicit Method

Δx	Δt	t	Maximum Error
0.0200	0.00833	0.1	0.692E $-$ 03
		0.2	0.110E $-$ 02
		0.3	0.134E $-$ 02
		0.4	0.148E $-$ 02
		0.5	0.157E $-$ 02
		0.6	0.162E $-$ 02
		0.7	0.165E $-$ 02
		0.8	0.167E $-$ 02
		0.9	0.168E $-$ 02
		1.0	0.169E $-$ 02
0.0200	0.0125	0.1	0.113E $-$ 02
		0.2	0.121E $-$ 01
		0.3	0.383E $+$ 00
		0.4	0.145E $+$ 02
		0.5	0.177E $+$ 05
		0.6	0.442E $+$ 75
		0.7	OVERFLOW
		0.8	OVERFLOW
		0.9	OVERFLOW
		1.0	OVERFLOW

and $q = Pf$, and the λ_i are the diagonal elements of D, that is, the eigenvalues of V. If all the eigenvalues λ_i are positive, then specifying all the z_i on the left endpoint is appropriate, and thus specification of u at $x = 0$ is justified.

The explicit upwind finite difference approximation 3.1.3 can be applied to 3.1.8, with the understanding that U, f, g, and h are vector functions, $v = V$ is now a matrix, and $a \equiv 0$. The error $e \equiv U - u$ then satisfies the following equation, derived by following the procedure used above in the single equation case. (T is the truncation error vector.)

$$\frac{e(x_i, t_{k+1}) - e(x_i, t_k)}{\Delta t} = -V \frac{e(x_i, t_k) - e(x_{i-1}, t_k)}{\Delta x} - T_i^k,$$

$$e(x_0, t_k) = 0,$$

$$e(x_i, t_0) = 0.$$

Replacing V by $P^{-1}DP$, and defining $\varepsilon \equiv Pe$, $\tau \equiv PT$, we have

$$\varepsilon(x_i, t_{k+1}) = [I - (\Delta t/\Delta x)D]\varepsilon(x_i, t_k) + (\Delta t/\Delta x)D\varepsilon(x_{i-1}, t_k) - \Delta t\, \tau_i^k,$$

$$\varepsilon(x_0, t_k) = 0,$$

$$\varepsilon(x_i, t_0) = 0.$$

Since D is a diagonal matrix, this vector equation separates out into m equations of the form 3.1.4. Thus, clearly, if $1 - (\Delta t/\Delta x)\lambda_{max}(V) \geq 0$, where $\lambda_{max}(V)$ is the largest of the (positive) eigenvalues of V, then all elements of the diagonal matrix $I - (\Delta t/\Delta x)\, D$ are positive, and the same analysis used to show stability for 3.1.4 can be used to show that each component of $\boldsymbol{\varepsilon}$ (and thus of $\mathbf{e}$) goes to zero as the truncation error goes to zero. The requirement for stability of the explicit upwind difference method applied to 3.1.8 is similar to 3.1.5, namely

$$\Delta t \leq \frac{\Delta x}{\lambda_{max}(V)}.$$

Clearly, if all eigenvalues of V are negative, then all unknowns should be specified on the right endpoint, and the explicit "upwind" (now biased toward $x = 1$) difference approximation is similarly conditionally stable. On the other hand, if l eigenvalues are positive and $m - l$ are negative, then l of the unknowns should be specified at $x = 0$ and the others at $x = 1$. In this case, some of the spatial derivatives must be biased toward $x = 0$ and others toward $x = 1$, and the stability analysis is more difficult.

3.2 The Method of Characteristics

The transport problem

$$u_t = -v(x, t)u_x + f(x, t, u),$$
$$u(0, t) = g(t), \qquad 0 \leq t \leq t_f, \qquad (3.2.1)$$
$$u(x, 0) = h(x), \qquad 0 \leq x \leq 1$$

(v is assumed to be positive), can also be solved by a technique called the "method of characteristics," which reduces the partial differential equation to an ordinary differential equation along the characteristic curves (cf. 3.1.6). Physically, a characteristic curve represents the trajectory that a given "particle" follows, transported by the convecting wind or current.

The parametric equations $(x, t) = (x(s), s)$ for the characteristic curve that starts at a point $(x_0, 0)$ on the initial line are found by solving

$$\frac{dx}{ds} = v(x(s), s),$$
$$x(0) = x_0. \qquad (3.2.2a)$$

Then, since $du/ds = u_x(dx/ds) + u_t(dt/ds) = u_x v + u_t = f$, the quasilinear problem 3.2.1 reduces to the following ordinary differential equation along this characteristic curve:

$$\frac{du}{ds} = f(x(s), s, u(s)),$$

$$u(0) = h(x_0).$$

(3.2.3a)

The parametric equations for the characteristic curve that starts at a point $(0, t_0)$ on the inlet boundary (see Figure 3.2.1) are found by solving

$$\frac{dx}{ds} = v(x(s), s),$$

$$x(t_0) = 0.$$

(3.2.2b)

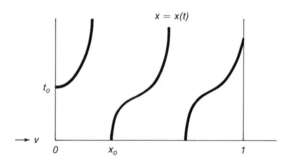

Figure 3.2.1

Along these characteristics, the problem 3.2.1 reduces to

$$\frac{du}{ds} = f(x(s), s, u(s)),$$

$$u(t_0) = g(t_0).$$

(3.2.3b)

To obtain a solution throughout the region $0 \le x \le 1, 0 \le t \le t_f$, we need to solve 3.2.2a and 3.2.2b for various values of x_0, t_0, thus obtaining the equations for various characteristics. It may be possible to solve 3.2.2 analytically—otherwise it is solved using a good initial value ordinary differential equation solver. Then along each characteristic, 3.2.3 is integrated, again by using an initial value solver.

As an example, 3.1.7 is solved again, this time by using the method of characteristics. Since $v = -2$, some characteristics are found by solving 3.2.2a:

$$\frac{dx}{ds} = -2,$$

$$x(0) = x_0,$$

which has the analytic solution $x(s) = x_0 - 2s$, so that the characteristics are the lines $x(t) = x_0 - 2t$. Along the characteristic $x(s) = x_0 - 2s$, the partial differential equation 3.1.7 reduces to the ordinary differential equation (cf. 3.2.3a)

$$\frac{du}{ds} = -u^2 + \exp[4s + 2(x_0 - 2s)] = -u^2 + \exp[2x_0],$$

$$u(0) = \exp[x_0].$$

This ordinary differential equation can also be solved analytically, yielding the constant solution

$$u(s) = \exp[x_0],$$

which agrees with the fact that along the characteristic that passes through $(x_0, 0)$, the known solution is

$$u(x, t) = \exp[2t + x] = \exp[2s + (x_0 - 2s)] = \exp[x_0].$$

The solutions along the characteristics that pass through $(1, t_0)$ are found similarly, by using 3.2.2b and 3.2.3b, modified to take into account the fact that the inlet boundary is now at $x = 1$, since v is negative.

Usually, of course, it will not be possible to solve the ordinary differential equations 3.2.2 and 3.2.3 analytically, but good, high accuracy, initial value solvers are available.

If v is a function of u, the two ordinary differential equations 3.2.2 and 3.2.3 are coupled and must be solved simultaneously, and the characteristic curves cannot be drawn a priori, before calculating u.

The method of characteristics can also be used to solve multi-dimensional problems. Consider, for example, the three-dimensional transport problem

$$u_t = -v_1(x, y, z, t, u)u_x - v_2(x, y, z, t, u)u_y$$
$$- v_3(x, y, z, t, u)u_z + f(x, y, z, t, u), \qquad (3.2.4)$$
$$u(x, y, z, 0) = h(x, y, z),$$
$$-\infty < x < \infty, \ -\infty < y < \infty, \ -\infty < z < \infty.$$

To find the solution $u(x(s), y(s), z(s), s)$ along the characteristic curve that begins at the point $(x_0, y_0, z_0, 0)$, we must solve a system of four coupled initial value ordinary differential equations (see Problem 6):

$$
\begin{aligned}
\frac{dx}{ds} &= v_1(x(s), y(s), z(s), s, u(s)), & x(0) &= x_0, \\[2mm]
\frac{dy}{ds} &= v_2(x(s), y(s), z(s), s, u(s)), & y(0) &= y_0, \\[2mm]
\frac{dz}{ds} &= v_3(x(s), y(s), z(s), s, u(s)), & z(0) &= z_0, \\[2mm]
\frac{du}{ds} &= f(x(s), y(s), z(s), s, u(s)), & u(0) &= h(x_0, y_0, z_0).
\end{aligned}
\tag{3.2.5}
$$

For problems with poorly behaved solutions, the method of characteristics is much more robust than finite differences. If the initial and/or boundary conditions are rough, this poses no problem for the method of characteristics, since the solution along each characteristic depends only on a single initial or boundary point, and in any case, very robust and accurate initial value solvers are available to solve the resulting ordinary differential equations.

On the other hand, even for a single hyperbolic equation, the method of characteristics is more difficult to apply. For a system of first-order equations, where there are generally several characteristics through a given point, the method of characteristics becomes much more cumbersome, and finite difference formulas are generally preferred.

3.3   An Explicit Method for the Wave Equation

Consider the one-dimensional vibrating string problem 3.0.2, normalized in the form

$$
\begin{aligned}
u_{tt} &= c^2 u_{xx} + f(x, t), \\
u(0, t) &= g_0(t), & 0 &\le t \le t_f, \\
u(1, t) &= g_1(t), \\
u(x, 0) &= h(x), & 0 &\le x \le 1, \\
u_t(x, 0) &= q(x),
\end{aligned}
\tag{3.3.1}
$$

where $c^2 = T/\rho$. A natural way to discretize 3.3.1 is to use the centered difference formula ($x_i = i \, \Delta x$, $t_k = k \, \Delta t$, $\Delta x = 1/N$):

$$\frac{U(x_i, t_{k+1}) - 2U(x_i, t_k) + U(x_i, t_{k-1})}{\Delta t^2}$$

$$= c^2 \frac{U(x_{i+1}, t_k) - 2U(x_i, t_k) + U(x_{i-1}, t_k)}{\Delta x^2} + f(x_i, t_k) \quad (3.3.2)$$

with

$$U(x_0, t_k) = g_0(t_k),$$

$$U(x_N, t_k) = g_1(t_k),$$

$$U(x_i, t_0) = h(x_i),$$

$$U(x_i, t_1) = h(x_i) + \Delta t \, q(x_i) + \tfrac{1}{2} \Delta t^2 [c^2 h''(x_i) + f(x_i, 0)].$$

It is clear from the stencil for this method (Figure 3.3.1) that it is explicit, since $U(x_i, t_{k+1})$ can be calculated explicitly from the values of U at the *two* previous time levels, t_k and t_{k-1}.

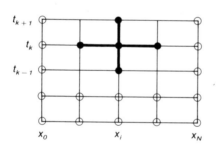

Figure 3.3.1

It is clear from this figure that we need to know U on the first *two* time levels (t_0 and t_1) in order to get the method started. (Points where U is known from the initial or boundary conditions are circled.) On level t_0, the solution values are known: $u(x, 0) = h(x)$. At $t = t_1 = \Delta t$ we have used the Taylor series expansion

$$u(x_i, \Delta t) = u(x_i, 0) + \Delta t \, u_t(x_i, 0) + \tfrac{1}{2} \Delta t^2 u_{tt}(x_i, 0) + O(\Delta t^3)$$

or

$$u(x_i, \Delta t) \approx h(x_i) + \Delta t \, q(x_i) + \tfrac{1}{2} \Delta t^2 [c^2 h''(x_i) + f(x_i, 0)]$$

to calculate approximate values $U(x_i, \Delta t)$, of $O(\Delta t^3)$ accuracy. The value of $u_{tt}(x_i, 0)$ was calculated using the partial differential equation 3.3.1. (If h'' is not available, a finite difference approximation may be substituted.) It will be seen later (formula 3.3.9) that retaining three terms in the Taylor series is necessary to preserve the second-order accuracy of this centered difference method.

As in Chapter 2, to study stability, we first note that the true solution satisfies

$$\frac{u(x_i, t_{k+1}) - 2u(x_i, t_k) + u(x_i, t_{k-1})}{\Delta t^2}$$

$$= c^2 \frac{u(x_{i+1}, t_k) - 2u(x_i, t_k) + u(x_{i-1}, t_k)}{\Delta x^2} + f(x_i, t_k) + T_i^k \quad (3.3.3)$$

with

$$u(x_0, t_k) = g_0(t_k),$$

$$u(x_N, t_k) = g_1(t_k),$$

$$u(x_i, t_0) = h(x_i),$$

$$u(x_i, t_1) = h(x_i) + \Delta t \, q(x_i) + \tfrac{1}{2} \Delta t^2 [c^2 h''(x_i) + f(x_i, 0)]$$

$$+ \tfrac{1}{6} \Delta t^3 u_{ttt}(x_i, \xi_i),$$

where the truncation error is (cf. 1.2.5b)

$$T_i^k = \frac{1}{12} \Delta t^2 u_{tttt}(x_i, \xi_{ik}) - \frac{c^2}{12} \Delta x^2 u_{xxxx}(\alpha_{ik}, t_k)$$

$$= O(\Delta t^2) + O(\Delta x^2).$$

Subtracting the difference equation satisfied by u (3.3.3) from the one satisfied by U (3.3.2) gives an equation for the error $e \equiv U - u$:

$$\frac{e(x_i, t_{k+1}) - 2e(x_i, t_k) + e(x_i, t_{k-1})}{\Delta t^2}$$

$$= c^2 \frac{e(x_{i+1}, t_k) - 2e(x_i, t_k) + e(x_{i-1}, t_k)}{\Delta x^2} - T_i^k \quad (3.3.4)$$

with

$$e(x_0, t_k) = 0,$$

$$e(x_N, t_k) = 0,$$

$$e(x_i, t_0) = 0,$$

$$e(x_i, t_1) = -\tfrac{1}{6} \Delta t^3 u_{ttt}(x_i, \xi_i).$$

The techniques used in the previous sections to bound the error will not work on this three-level scheme. So we will use the Fourier stability method (discussed further in Appendix 1) to actually solve the difference equation 3.3.4 for the error.

For a fixed t_k, the error can be thought of as a vector (of length $N - 1$, since $e(x_0, t_k) = e(x_N, t_k) = 0$). Since the $N - 1$ vectors

$$(\sin(m\pi x_1), \sin(m\pi x_2), \ldots, \sin(m\pi x_{N-1})), \qquad m = 1, \ldots, N - 1$$

are shown in Problem 7a to be linearly independent, the error can be expressed as a linear combination of these basis vectors:

$$e(x_i, t_k) = \sum_{m=1}^{N-1} a_m(t_k) \sin(m\pi x_i). \tag{3.3.5}$$

Note that $e(x_0, t_k) = e(x_N, t_k) = 0$, regardless of how the coefficients are chosen. We also expand the truncation error (for fixed k) in this basis:

$$T_i^k = \sum_{m=1}^{N-1} b_m(k) \sin(m\pi x_i). \tag{3.3.6}$$

If these expressions for e and T_i^k are plugged into 3.3.4, after equating coefficients of $\sin(m\pi x_i)$, we get (using the trigonometric identity $\sin(\alpha + \beta) = \sin(\alpha)\cos(\beta) + \cos(\alpha)\sin(\beta)$ to simplify)

$$\frac{a_m(t_{k+1}) - 2a_m(t_k) + a_m(t_{k-1})}{\Delta t^2} = c^2 \frac{2\cos(m\pi\,\Delta x) - 2}{\Delta x^2} a_m(t_k) - b_m(k)$$

or

$$a_m(t_{k+1}) + [-2 + 4c^2(\Delta t^2/\Delta x^2)\sin^2(\tfrac{1}{2}m\pi\,\Delta x)]a_m(t_k) + a_m(t_{k-1})$$
$$= -\Delta t^2 b_m(k).$$

Now Theorem 1.3.1 can be applied to this difference equation (with m considered to be fixed). If $c^2\,\Delta t^2/\Delta x^2 \leq 1$, that is, if

$$\Delta t \leq \frac{\Delta x}{c}, \tag{3.3.7}$$

then, for all m, the discriminant of the characteristic polynomial

$$\lambda^2 + \left[-2 + 4c^2\left(\frac{\Delta t^2}{\Delta x^2}\right)\sin^2\left(\tfrac{1}{2}m\pi\,\Delta x\right)\right]\lambda + 1$$

is negative, so the two roots are a complex conjugate pair. Each of these simple roots has absolute value equal to one, so by Theorem 1.3.1,

$$|a_m(t_k)| \leq M_\rho \left[\max\{|a_m(t_0)|, |a_m(t_1)|\} + t_k \, \Delta t \max_{j \leq k} |b_m(j)| \right]. \quad (3.3.8)$$

This bounds the Fourier coefficients of the error in terms of the coefficients of the truncation and initial errors. In order to translate this into a bound on the error itself, we use the (discrete) inverse Fourier transforms of the forward transforms 3.3.5 and 3.3.6 (see Problem 7b) to expand $b_m(j)$, $a_m(t_0)$ and $a_m(t_1)$:

$$b_l(j) = \frac{2}{N} \sum_{i=1}^{N-1} T_i^j \sin(l\pi x_i),$$

$$a_l(t_j) = \frac{2}{N} \sum_{i=1}^{N-1} e(x_i, t_j) \sin(l\pi x_i).$$

By defining $T_{\max} \equiv \max |T_i^k|$, and by using E_k (as before) to denote the maximum error at the time level t_k, we have

$$|b_l(j)| \leq \left(\frac{2}{N}\right)(N-1)T_{\max} \leq 2T_{\max},$$

$$|a_l(t_0)| \leq \left(\frac{2}{N}\right)(N-1)E_0 = 0,$$

$$|a_l(t_1)| \leq \left(\frac{2}{N}\right)(N-1)E_1 \leq 2E_1.$$

So (3.3.8) becomes

$$|a_m(t_k)| \leq M_\rho[2E_1 + t_k \, \Delta t \, (2T_{\max})].$$

Then, from (3.3.5), it is seen that

$$|e(x_i, t_k)| \leq (N-1) \max_m |a_m(t_k)|$$

$$\leq \left(\frac{\Delta t}{\Delta x}\right) M_\rho \left[\frac{2E_1}{\Delta t} + 2t_k T_{\max}\right],$$

so

$$E_k \leq \frac{M_\rho}{c} \left[\frac{2E_1}{\Delta t} + 2t_k T_{\max}\right], \quad (3.3.9)$$

where $\Delta t \leq \Delta x/c$ (formula 3.3.7) was used to obtain this last bound.

Since the maximum error at time level t_1 is $O(\Delta t^3)$ (see 3.3.4), the explicit method 3.3.2 is stable, and the error is $O(\Delta x^2) + O(\Delta t^2)$, under the assumption that $\Delta t \leq \Delta x/c$.

That this method is unstable when 3.3.7 is violated, that is, when $\Delta t > \Delta x/c$, can be seen by looking at the domain of dependence of the approximate solutions. If Δt and Δx go to zero with a fixed ratio $\Delta t/\Delta x = r$, the value of the approximate solution at a fixed point (x, t) will depend only on the initial values between $x - t/r$ and $x + t/r$ (see Figure 3.3.2). Thus the limit of the approximate solutions also depends only on the initial values in $(x - t/r, x + t/r)$.

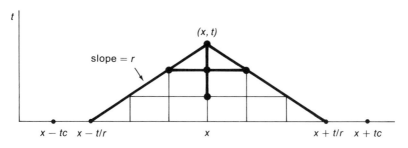

Figure 3.3.2

The true solution at (x, t), however, depends on the initial values in the interval $(x - ct, x + ct)$, as seen by looking at the d'Alembert solution 3.0.5 to this problem (with $f = 0$). If $c > 1/r = \Delta x/\Delta t$, or $\Delta t > \Delta x/c$, then the approximate solutions cannot possibly be converging to the true solution, because we can change the initial values outside $(x - t/r, x + t/r)$ but still inside $(x - ct, x + ct)$, changing the true solution but not the limit of the approximate solutions. Thus, for stability we must have $\Delta t \leq \Delta x/c$, as determined earlier (3.3.7).

Since $\Delta t \leq \Delta x/c$ is not a very restrictive requirement (Δt should be $O(\Delta x)$ for accuracy considerations anyway), implicit methods are not very attractive for the wave problem, especially in two or more space dimensions where a very large linear system would have to be solved each time step. Unconditionally stable implicit methods for the wave equation do exist, however (see Appendix 1).

3.4 A Damped Wave Example

As an example, we will solve a two-dimensional "damped" wave problem (cf. 3.0.4) in a triangular region:

$$u_{tt} + u_t = u_{xx} + u_{yy} \qquad \text{in } R,$$

$$u = 0 \qquad \text{on the boundary,}$$

$$u = xy(x + y - 1) \qquad \text{at } t = 0,$$

$$u_t = 0 \qquad \text{at } t = 0.$$

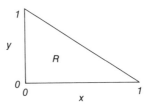

This models a (damped) triangular elastic membrane that is fastened along the edges and released from a loaded position at $t = 0$. The obvious generalization of 3.3.2 is used to approximate this partial differential equation:

$$\frac{U(x_i, y_j, t_{k+1}) - 2U(x_i, y_j, t_k) + U(x_i, y_j, t_{k-1})}{\Delta t^2}$$

$$+ \frac{U(x_i, y_j, t_{k+1}) - U(x_i, y_j, t_{k-1})}{2\Delta t}$$

$$= \frac{U(x_{i+1}, y_j, t_k) - 2U(x_i, y_j, t_k) + U(x_{i-1}, y_j, t_k)}{\Delta x^2}$$

$$+ \frac{U(x_i, y_{j+1}, t_k) - 2U(x_i, y_j, t_k) + U(x_i, y_j, y_{j-1}, t_k)}{\Delta y^2}$$

$$(3.4.1)$$

with

$$U(x_i, y_j, t_0) = x_i y_j (x_i + y_j - 1),$$
$$U(x_i, y_j, t_1) = x_i y_j (x_i + y_j - 1) + \Delta t^2 (x_i + y_j),$$

and

$$U(x_i, y_j, t_k) = 0 \qquad \text{when } (x_i, y_j) \text{ is on the boundary.}$$

The values of u at $t = t_1$ were estimated using the Taylor series

$$u(x, y, \Delta t) \approx u(x, y, 0) + \Delta t \, u_t(x, y, 0) + \tfrac{1}{2}\Delta t^2 u_{tt}(x, y, 0)$$
$$= u(x, y, 0) + 0$$
$$+ \tfrac{1}{2}\Delta t^2 [u_{xx}(x, y, 0) + u_{yy}(x, y, 0) - u_t(x, y, 0)]$$
$$= xy(x + y - 1) + \tfrac{1}{2}\Delta t^2 [2y + 2x].$$

If $\Delta x \neq \Delta y$, the diagonal boundary presents a big headache for the finite difference method, as there may exist interior grid points (x_i, y_j) that are connected through the finite difference formula with points outside the region. Therefore, we use $\Delta x = \Delta y$, so that the neighbors of any interior grid points are all either interior or exactly on the boundary (see Figure 3.4.1). If the region has a curved boundary segment, it is still possible to treat it with finite differences, but finite elements (see Chapter 5) are much better suited for such regions.

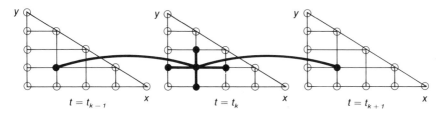

$t = t_{k-1}$ $t = t_k$ $t = t_{k+1}$

Figure 3.4.1

Since 3.4.1 is explicit, the program to implement this finite difference method is very simple and is shown in Figure 3.4.2. With $\Delta x = \Delta y = 0.025$, $\Delta t = 0.01$, the solution at $(0.25, 0.25)$, as a function of time, is shown in Figure 3.4.3, along with that calculated by **PDE/PROTRAN**, a finite element program discussed in Section 5.6. Note the damping with time of the amplitude of the oscillations of the vibrating membrane.

```
      IMPLICIT DOUBLE PRECISION(A-H,O-Z)
C                               N = NUMBER OF POINTS IN X,Y DIRECTIONS
C                               M = NUMBER OF TIME STEPS
      PARAMETER (N=40, M=200)
      DIMENSION UKM1(0:N,0:N),UK(0:N,0:N),UKP1(0:N,0:N)
      DX = 1.D0/N
      DY = DX
      DT = 2.D0/M
C                               BOUNDARY AND INTERIOR VALUES SET TO ZERO
C                               (BOUNDARY VALUES WILL NEVER CHANGE)
      DO 5 I=0,N
      DO 5 J=0,N
        UKM1(I,J) = 0.0
        UK  (I,J) = 0.0
        UKP1(I,J) = 0.0
    5 CONTINUE
C                               RESET VALUES OF UKM1,UK AT INTERIOR POINTS
C                               TO INITIAL VALUES AT T0,T1
      DO 10 I=1,N-2
      DO 10 J=1,N-I-1
        XI = I*DX
        YJ = J*DY
        UKM1(I,J) = XI*YJ*(XI+YJ-1)
        UK  (I,J) = UKM1(I,J) + DT*DT*(XI+YJ)
   10 CONTINUE
C                               BEGIN MARCHING FORWARD IN TIME
      DO 30 K=1,M
        DO 15 I=1,N-2
        DO 15 J=1,N-I-1
          UKP1(I,J) = ( 2*UK(I,J) + (-1+DT/2)*UKM1(I,J)
     &      +(UK(I+1,J)-2*UK(I,J)+UK(I-1,J))*DT**2/DX**2
     &      +(UK(I,J+1)-2*UK(I,J)+UK(I,J-1))*DT**2/DY**2 )/ (1+DT/2)
   15   CONTINUE
C                               UPDATE UKM1,UK
        DO 20 I=0,N
        DO 20 J=0,N
          UKM1(I,J) = UK(I,J)
          UK  (I,J) = UKP1(I,J)
   20   CONTINUE
        TKP1 = (K+1)*DT
C                               OUTPUT SOLUTION AT X=0.25,Y=0.25
        PRINT 25, TKP1,UKP1(N/4,N/4)
   25   FORMAT (2E15.5)
   30 CONTINUE
      STOP
      END
```

Figure 3.4.2
FORTRAN77 Program to Solve Elastic Membrane Problem

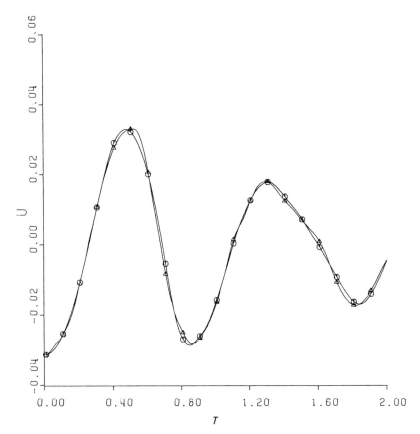

Figure 3.4.3
Displacement at (0.25, 0.25) for Elastic Membrane Problem
○ = finite difference solution
△ = finite element (PDE/PROTRAN) solution

With $\Delta x = \Delta y = 0.025$ and $\Delta t = 0.02$, the explicit method experienced instability. The stability condition for this two-dimensional wave equation is shown in Appendix 1 to be $\Delta t \leq \Delta x / \sqrt{2}$, which is not satisfied for these values of Δx, Δy, Δt.

The methods discussed in this chapter work well as long as the solution is well behaved, but unfortunately this is often not the case. For this reason, hyperbolic problems are probably the most difficult, as a class, of the differential equations we will study, despite our success with these simple examples.

3.5 Problems

1. a. Verify that if u_1 and u_2 are two solutions of the transport equation (cf. 3.0.1)

$$u_t = -\mathbf{V}^T(u\mathbf{v}) + f(x, y, z, t) \qquad \text{in } R,$$

$$u = g(x, y, z, t) \qquad \text{on the part of the boundary where } \mathbf{v}^T\mathbf{n} < 0,$$

$$u = h(x, y, z) \qquad \text{when } t = 0,$$

where $\mathbf{v}$ is a constant vector, then their difference, $e \equiv u_1 - u_2$, satisfies

$$(e^2)_t = -\mathbf{V}^T(e^2\mathbf{v}).$$

Using this result, show that there is at most one (smooth) solution to the above transport problem.

Hint: Show that the integral over R of e^2 is nonincreasing and is zero when $t = 0$.

   b. Verify that if u_1 and u_2 are two solutions of the damped wave equation (cf. 3.0.4)

$$u_{tt} = -au_t + c^2\mathbf{V}^2 u + f(x, y, z, t) \qquad \text{in } R,$$

$$u = g_0(x, y, z, t) \qquad \text{on part of the boundary,}$$

$$\partial u/\partial n = g_1(x, y, z, t) \qquad \text{on the other part,}$$

$$u = h(x, y, z) \qquad \text{when } t = 0,$$

$$u_t = q(x, y, z) \qquad \text{when } t = 0,$$

where $a \geq 0$ and $c > 0$ are constant, then their difference, $e \equiv u_1 - u_2$, satisfies

$$[(e_t)^2 + c^2|\mathbf{V}e|^2]_t = 2\,\mathbf{V}^T(c^2 e_t\,\mathbf{V}e) - 2a(e_t)^2.$$

Using this result, show that there is at most one (smooth) solution to the above wave problem.

Hint: Show that the integral over R of $(e_t)^2 + c^2|\mathbf{V}e|^2$ is nonincreasing, and is zero when $t = 0$.

2. Verify that the d'Alembert solution 3.0.5 is indeed the solution of the wave equation:

$$u_{tt} = c^2 u_{xx},$$

$$u(x, 0) = h(x), \qquad -\infty < x < \infty,$$

$$u_t(x, 0) = q(x).$$

3. Verify that using an "upwind" scheme, where the approximation to u_x is biased toward $x = 0$ if $v > 0$ (formula 3.1.3) and toward $x = 1$ if $v < 0$, to approximate 3.1.1 is equivalent to using the centered difference scheme 2.1.3 to approximate the diffusion-convection problem 2.1.1, when $D = \frac{1}{2}|v| \, \Delta x$. Consider the two cases $v > 0$ and $v < 0$ separately.

4. Consider the first-order transport equation

$$u_t = -2t(1 + x)u_x + \ln(1 + x) + 2u,$$

$$u(x, 0) = 0,$$

$$u(0, t) = 1.$$

   a. Find the form of the characteristic curves by obtaining the general solution of the differential equation 3.2.2, which can be easily solved analytically.
   b. Above the characteristic curve that passes through $(0, 0)$, the solution depends only on the boundary conditions, and below this curve the solution depends only on the initial conditions. Since the initial and boundary conditions do not match at $(0, 0)$, the solution will have a discontinuity along this characteristic. Find the point $(1, T_c)$ where this characteristic intersects the line $x = 1$.
   c. Find the point $(x_0, 0)$ where the characteristic through $(1, \frac{1}{2})$ intersects the line on which the initial conditions are given. Then find $u(1, \frac{1}{2})$ by integrating (analytically) the initial value problem 3.2.3a along this characteristic. (Answer: $u(1, \frac{1}{2}) = 0.43529$.)
   d. Find the point $(0, t_0)$ where the characteristic through $(1, 1)$ intersects the line on which the boundary conditions are given. Then find $u(1, 1)$ by integrating (analytically) the initial value problem 3.2.3b along this characteristic. (Answer: $u(1, 1) = 2.62969$.)

5. Solve the first-order transport equation of Problem 4 with the upwind finite difference method 3.1.3, and compare your computed solution at the points $(1, \frac{1}{2})$ and $(1, 1)$ with those found using the method of characteristics. Does the discontinuity in the solution seem to create any difficulties for the finite difference method?

6. Verify that if $x(s), y(s), z(s), u(s)$ is the solution to the ordinary differential equation system 3.2.5, then the value of the solution of 3.2.4 at the point $(x(s), y(s), z(s), s)$ is $u(s)$.

7. a. Show that $(x_i = i\,\Delta x,\ \Delta x = 1/N)$

$$\sum_{i=1}^{N-1} \sin(m\pi x_i)\sin(l\pi x_i) = \begin{cases} 0, & \text{if } m \neq l, \\ N/2 & \text{if } m = l, \end{cases}$$

so that vectors $(\sin(m\pi x_1), \ldots,\ \sin(m\pi x_{N-1}))$, $m = 1, \ldots,\ N-1$ are mutually orthogonal and therefore linearly independent.

Hint: Use $\sin\theta = [e^{I\theta} - e^{-I\theta}]/(2I)$, where $I \equiv \sqrt{-1}$
and $1 + r + r^2 + r^3 + \cdots + r^{N-1} = [1 - r^N]/[1 - r]$.

b. Derive the inverse Fourier transform formula used in Section 3.3. That is, show that if

$$e_i = \sum_{m=1}^{N-1} a_m \sin(m\pi x_i),$$

then

$$a_l = \frac{2}{N} \sum_{i=1}^{N-1} e_i \sin(l\pi x_i).$$

Hint: Multiply both sides of the top equation by $\sin(l\pi x_i)$, and sum from $i = 1$ to $i = N - 1$. Then use Problem 7a to evaluate these sums.

8. Solve the one-dimensional wave equation

$$u_{tt} = u_{xx} + u + 2e^t,$$

$$u(0, t) = 0,$$

$$u(1, t) = 0,$$

$$u(x, 0) = x(1 - x),$$

$$u_t(x, 0) = x(1 - x),$$

by using centered finite difference formulas (cf. 3.3.2). On the second time level, use

a. $U(x_i, t_1) = u(x_i, 0) + \Delta t\, u_t(x_i, 0) + \frac{1}{2}\,\Delta t^2 u_{tt}(x_i, 0)$

$$= u(x_i, 0) + \Delta t\, u_t(x_i, 0) + \frac{1}{2}\,\Delta t^2 [u_{xx}(x_i, 0) + u(x_i, 0) + 2].$$

b. $U(x_i, t_1) = u(x_i, 0) + \Delta t\, u_t(x_i, 0).$

In each case, calculate the maximum error at $t = 1$ when $\Delta x = \Delta t = 0.02$ and $\Delta x = \Delta t = 0.01$. Since the exact solution is $u(x, t) = e^t x(1 - x)$, the error depends only on Δt, not Δx. (Why?) Thus,

$\log_2(\text{error}_{0.02}/\text{error}_{0.01})$ is an experimental estimate of the exponent of Δt in the error. Calculate this exponent and verify that when three terms in the Taylor series expansion for $U(x_i, t_1)$ are used (a), the error is $O(\Delta t^2)$, while if only two terms are used (b), the error is only $O(\Delta t)$.

Also do one experiment using three terms (a), with Δt chosen to slightly violate the stability condition $\Delta t \leq \Delta x/c$ (formula 3.3.7).

9. Use the Fourier stability method outlined in Appendix 1 to determine the conditions under which the following approximations to $u_t = u_x$ and $u_{tt} = u_{xx}$, respectively, are stable.

a. $\dfrac{U(x_i, t_{k+1}) - U(x_i, t_{k-1})}{2\,\Delta t} = \dfrac{U(x_{i+1}, t_k) - U(x_{i-1}, t_k)}{2\,\Delta x},$

b. $\dfrac{U(x_i, t_{k+1}) - 2U(x_i, t_k) + U(x_i, t_{k-1})}{\Delta t^2}$

$\qquad = \dfrac{U(x_{i+1}, t_{k+1}) - 2U(x_i, t_{k+1}) + U(x_{i-1}, t_{k+1})}{\Delta x^2}.$

4

Boundary Value Problems

4.0 Introduction

Boundary value problems generally model steady-state (time independent) phenomena. For example, if we wish to find the steady-state distribution of a diffusing element, we set the left-hand side of the diffusion-convection equation 2.0.3 equal to zero, since "steady-state" implies that the solution is not changing with time, so that $u_t = 0$. This gives the steady-state diffusion-convection equation

$$0 = \mathbf{V}^T(D\mathbf{V}\mathbf{u}) - \mathbf{V}^T(u\mathbf{v}) + f \tag{4.0.1}$$

or

$$0 = (Du_x)_x + (Du_y)_y + (Du_z)_z - (uv_1)_x - (uv_2)_y - (uv_3)_z + f$$

where v_1, v_2, v_3, D, and f may depend on x, y, z, and u but not on t—otherwise there is no steady-state. Boundary conditions for the steady-state problem may be the same as for the time-dependent problem; and usually either the concentration (temperature, if heat is what is diffusing) u, or the boundary flux $\mathbf{J}^T\mathbf{n} = -D(\partial u/\partial n) + u\mathbf{v}^T\mathbf{n}$ is given on each portion of the boundary. The initial conditions no longer influence the concentration distribution once a steady-state is reached, so they are not present in the boundary value formulation.

The damped wave and elastic body equations 3.0.4 and 3.0.7 also generate boundary value problems when a steady-state is assumed.

Another boundary value problem can be derived by considering potential energy in any inverse square force field. If there is only a single point source for this field, of strength s at the point $(\alpha \, \beta, \Gamma)$, then the potential energy at any other point (x, y, z) due to this source is

$$u(x, y, z) = \frac{-s}{\sqrt{(x - \alpha)^2 + (y - \beta)^2 + (z - \Gamma)^2}} \equiv \frac{-s}{r}.$$

It is easy to verify that

$$\mathbf{\nabla u} = \frac{s}{r^2} \left[\frac{(x - \alpha)}{r}, \frac{(y - \beta)}{r}, \frac{(z - \Gamma)}{r} \right]$$

so that, since force is equal to the negative of the gradient of the potential energy, the force field at (x, y, z) is in the direction of (α, β, Γ) and of magnitude s/r^2. If the field is the result of a continuum of sources, the potential at a point (x, y, z) away from the sources is

$$u(x, y, z) = \iiint\limits_{S} \frac{-s(\alpha, \beta, \Gamma) \, d\alpha \, d\beta \, d\Gamma}{\sqrt{[(x - \alpha)^2 + (y - \beta)^2 + (z - \Gamma)^2]}}. \qquad (4.0.2)$$

Here S is the set of all source points (assumed not to include (x, y, z)), $s(\alpha, \beta, \Gamma)$ is the strength per unit volume of the source at (α, β, Γ), and the potential at (x, y, z) is the sum of the potentials due to all the sources.

From 4.0.2 it can be verified by direct calculation (Problem 2) that the potential energy satisfies $\nabla^2 u = 0$, or $u_{xx} + u_{yy} + u_{zz} = 0$, called Laplace's equation. Boundary conditions are generally similar to those for the diffusion equation.

It should be mentioned that the solution to Laplace's equation at any point (x, y, z) can be shown to be exactly equal to the average of the values of the solution on the surface of any sphere with center (x, y, z), assuming that Laplace's equation holds throughout the sphere. Thus the solution to Laplace's equation never has a relative maximum or a relative minimum in the interior of the solution domain. For if it had, say, a maximum at (x, y, z), its value there would be greater than all of its values on the surface of any sufficiently small sphere with center (x, y, z), thus greater than its average on that sphere. This "maximum principle" is reasonable if the solution is thought of as representing the steady-state density distribution of a diffusing element (without sources or convection) or the height of an (unloaded) elastic membrane. In neither case would we expect a local maximum or minimum to occur in the interior.

While initial value problems can be solved by marching forward in time from the initial values, we almost always have to solve a system of simultaneous algebraic equations in order to solve a boundary value problem. Since consistent but unstable methods for boundary value problems are encountered far less frequently than for initial value problems, the principal challenge is no longer to avoid instability, but to solve this linear or nonlinear algebraic system, which may involve a large number of equations and unknowns.

An *eigenvalue* boundary value problem generally consists of a linear homogeneous differential equation, with homogeneous boundary conditions, that contains an undetermined parameter. The problem is to determine values of this parameter for which the boundary problem has a nonzero solution (zero is a solution of any homogeneous boundary value problem). Such values are called the eigenvalues, and the corresponding solutions are the eigenfunctions.

To illustrate the physical situations in which eigenvalue differential equation problems may arise, consider the vibrating string problem 3.3.1. Suppose that the extremes of the string are fixed so that $g_0(t) = g_1(t) = 0$, and a loading function is applied that is periodic in time and has the form $f(x, t) = F(x) \sin(\omega t)$. Then, although the model 3.3.1 does not take this into account, a real string will always experience some slight damping, so that eventually the effects of the initial conditions will die out, and it is physically evident that the solution will ultimately be periodic in time with the same frequency and phase as the applied load, so that $u(x, t) = A(x) \sin(\omega t)$. Substituting this expression for the solution into the partial differential equation 3.3.1 gives an ordinary differential equation boundary value problem for the amplitude of the vibrations:

$$-\omega^2 A(x) = c^2 A''(x) + F(x),$$

$$A(0) = 0, \qquad\qquad\qquad (4.0.3)$$

$$A(1) = 0.$$

This nonhomogeneous problem can be solved by first finding the general solution of the homogeneous version of 4.0.3, $-c^2 A''(x) = \omega^2 A(x)$, namely $A(x) = R \sin(\omega x/c) + S \cos(\omega x/c)$. A particular solution of the differential equation in 4.0.3 may now be found by using the well-known "variation of parameters" technique, which yields (see Problem 2)

$$A_p(x) = P(x) \sin\left(\frac{\omega x}{c}\right) + Q(x) \cos\left(\frac{\omega x}{c}\right), \qquad (4.0.4)$$

where

$$P(x) = \frac{-1}{(c\omega)} \int_0^x \cos\left(\frac{\omega s}{c}\right) F(s) \, ds,$$

$$Q(x) = \frac{1}{(c\omega)} \int_0^x \sin\left(\frac{\omega s}{c}\right) F(s) \, ds.$$

The general solution of the nonhomogeneous differential equation is $A(x) = A_p(x) + R \sin(\omega x/c) + S \cos(\omega x/c)$. Applying the boundary conditions reveals the solution of 4.0.3 to be

$$A(x) = A_p(x) - A_p(1) \frac{\sin(\omega x/c)}{\sin(\omega/c)}.$$

From this equation it is seen that the amplitude of the periodic string vibration becomes large without bounds as ω/c approaches an integer multiple of π, that is, as $\omega \to k\pi c$. When ω is exactly equal to $k\pi c$, the homogeneous version of 4.0.3,

$$-c^2 A''(x) = \omega^2 A(x),$$

$$A(0) = 0, \qquad (4.0.5)$$

$$A(1) = 0,$$

has a nonunique (nonzero) solution, $A(x) = R \sin(k\pi x)$, where R is arbitrary, so the nonhomogeneous problem 4.0.3 either has no solution or many.

We say that $\pi c, 2\pi c, 3\pi c, \ldots$ are the "resonant frequencies" of the string and that the eigenvalues of the *eigenvalue problem* 4.0.5 are $(\pi c)^2$, $(2\pi c)^2$, $(3\pi c)^2, \ldots$. Of course, in a real experiment, there will be some slight energy dissipation (damping), so that when a loading is applied that has a frequency equal to one of the resonant frequencies of the string, the amplitude will become large, but not infinite.

4.1 Finite Difference Methods

The steady-state version of the linear, one-dimensional diffusion–convection problem 2.1.1 is

$$0 = D(x)u_{xx} - v(x)u_x + a(x)u + f(x),$$

$$u(0) = g_0,$$

$$u(1) = g_1. \tag{4.1.1}$$

It is assumed that $D > 0$.

This is actually an ordinary differential equation. If u_x and u_{xx} are approximated by the finite difference formulas 1.2.5a–b, we have the "steady-state" version of the finite difference equations 2.1.3, which has truncation error $O(\Delta x^2)$:

$$0 = D(x_i) \frac{U(x_{i+1}) - 2U(x_i) + U(x_{i-1})}{\Delta x^2}$$

$$- v(x_i) \frac{U(x_{i+1}) - U(x_{i-1})}{2\,\Delta x} + a(x_i)U(x_i) + f(x_i) \tag{4.1.2}$$

$$U(x_0) = g_0,$$

$$U(x_N) = g_1.$$

Here, as in Section 2.1, $x_i = i\,\Delta x$ and $\Delta x = 1/N$.

Equations 4.1.2 represent a system of $N - 1$ linear equations for the $N - 1$ unknowns $U(x_i)$, $i = 1, \ldots, N - 1$. This system is tridiagonal, since each equation involves only three unknowns:

$$\left[\frac{-D(x_i)}{\Delta x^2} - \frac{\frac{1}{2}v(x_i)}{\Delta x} \right] U(x_{i-1}) + \left[\frac{2D(x_i)}{\Delta x^2} - a(x_i) \right] U(x_i) \tag{4.1.3}$$

$$+ \left[\frac{-D(x_i)}{\Delta x^2} + \frac{\frac{1}{2}v(x_i)}{\Delta x} \right] U(x_{i+1}) = f(x_i).$$

If $a \le 0$, and Δx is small enough so that $\Delta x\, V_{\max} < 2D_{\min}$ ($V_{\max} \equiv \max |v(x)|$ and $D_{\min} \equiv \min D(x)$), this tridiagonal system has a unique solution. To show that the linear system is nonsingular, it is sufficient to show that the homogeneous system ($f = g_0 = g_1 = 0$) has only the zero solution. Thus let us assume the contrary, namely that the homogeneous problem has a solution $U(x_i)$ which is *not* identically zero. Let S be the set of grid points where $|U|$ attains its (positive) maximum value. Since not all grid points belong to S (at least the boundary points do not), there must be an (interior) point x_l in S that has at least one neighbor not in S. Then, since $f = 0$, 4.1.3 gives

$$\left\{\frac{2D(x_l)}{\Delta x^2} - a(x_l)\right\}|U(x_l)|$$

$$\leq \left|\frac{D(x_l)}{\Delta x^2} + \frac{\frac{1}{2}v(x_l)}{\Delta x}\right| \cdot |U(x_{l-1})| + \left|\frac{D(x_l)}{\Delta x^2} - \frac{\frac{1}{2}v(x_l)}{\Delta x}\right| \cdot |U(x_{l+1})| \quad (4.1.4)$$

$$< \left\{\frac{2D(x_l)}{\Delta x^2}\right\}|U(x_l)|,$$

where the last inequality follows from the fact that $D(x_l)/\Delta x^2 \pm \frac{1}{2}v(x_l)/\Delta x$ is positive, and the facts that

$$|U(x_{l-1})| \leq |U(x_l)| \qquad \text{and} \qquad |U(x_{l+1})| \leq |U(x_l)|,$$

with strict inequality holding in at least one case. But 4.1.4 is impossible, since $a \leq 0$; the assumption that the homogeneous problem has a nonzero solution must be false, and therefore 4.1.3 represents a nonsingular linear system.

Under the above assumptions on $a(x)$ and Δx, we can also see that the system 4.1.3 is weakly diagonal dominant. For (when $2 \leq i \leq N - 2$),

$$\left|\frac{-D(x_i)}{\Delta x^2} - \frac{\frac{1}{2}v(x_i)}{\Delta x}\right| + \left|\frac{-D(x_i)}{\Delta x^2} + \frac{\frac{1}{2}v(x_i)}{\Delta x}\right|$$

$$= \frac{D(x_i)}{\Delta x^2} + \frac{\frac{1}{2}v(x_i)}{\Delta x} + \frac{D(x_i)}{\Delta x^2} - \frac{\frac{1}{2}v(x_i)}{\Delta x}$$

$$= \frac{2D(x_i)}{\Delta x^2} \leq \left|\frac{2D(x_i)}{\Delta x^2} - a(x_i)\right|.$$

For $i = 1$ or $i = N - 1$, only one of the off-diagonal terms is present, so the diagonal is dominant on the first and last rows also.

As was shown in Theorem 0.2.2, this ensures that pivoting is not required during Gaussian elimination, since the matrix is nonsingular. A tridiagonal solver such as the one shown in Figure 0.4.4 can therefore be used.

In Problem 4 it is shown that the finite difference method 4.1.2 is stable and that the error has the same order $O(\Delta x^2)$ as the truncation error, when $a(x)$ is negative.

On the other hand, if $a(x)$ is positive for some values of x, the boundary value problem itself may not have a unique solution. For example, the problem

$$u_{xx} + \pi^2 u = 0, \qquad u(0) = u(1) = 0,$$

has $u(x) = R \sin(\pi x)$ as a solution for any value of R.

However, even when $a(x)$ is positive, if the boundary value problem *does* have a unique solution, we can still attempt to solve the finite difference equations 4.1.2. In fact, if Δx is sufficiently small, none of the diagonal elements are zero before the elimination begins, and a subroutine that does no pivoting (e.g., Figure 0.4.4) will fail only if during the course of the elimination a pivot becomes zero or nearly zero. Nevertheless, since the work and storage required are still $O(N)$, partial pivoting (Figure 0.4.6) is recommended for the sake of safety.

4.2 A Nonlinear Example

Consider the nonlinear problem

$$u_{xx} - u_x^2 - u^2 + u + 1 = 0,$$
$$u(0) = 0.5, \tag{4.2.1}$$
$$u(\pi) = -0.5,$$

which has an exact solution $u(x) = \sin(x + \pi/6)$. The corresponding centered finite difference equations are nonlinear ($x_i = i\,\Delta x$, $\Delta x = \pi/N$):

$$f_i \equiv \frac{U(x_{i+1}) - 2U(x_i) + U(x_{i-1})}{\Delta x^2} - \left[\frac{U(x_{i+1}) - U(x_{i-1})}{2\,\Delta x}\right]^2$$

$$- U(x_i)^2 + U(x_i) + 1 = 0, \tag{4.2.2}$$

$$U(x_0) = 0.5,$$
$$U(x_N) = -0.5.$$

Newton's method will be used to solve this nonlinear system, so the Jacobian matrix elements ($J_{ij} = \partial f_i/\partial U(x_j)$) are required:

$$J_{i,i} = \frac{-2}{\Delta x^2} - 2U(x_i) + 1,$$

$$J_{i,i-1} = \frac{1}{\Delta x^2} + \frac{U(x_{i+1}) - U(x_{i-1})}{2\,\Delta x^2}, \tag{4.2.3}$$

$$J_{i,i+1} = \frac{1}{\Delta x^2} - \frac{U(x_{i+1}) - U(x_{i-1})}{2\,\Delta x^2}.$$

All other elements in the ith row of the Jacobian are zero, hence the Jacobian matrix is tridiagonal.

Newton's method is defined by the iteration

$$\mathbf{U^{n+1} = U^n - \Delta^n} \qquad (4.2.4)$$

where

$$J^n \mathbf{\Delta^n = f^n}$$

and

$$\mathbf{U^n} \equiv (U^n(x_1), \ldots, U^n(x_{N-1})),$$

and the components of the vector $\mathbf{f^n}$ and its Jacobian matrix J^n are defined by 4.2.2 and 4.2.3. The superscript n means that the function components and Jacobian elements are evaluated at $\mathbf{U^n}$.

The results of these calculations are shown in Table 4.2.1. When a starting guess of $U^0(x_i) = 0$ was used for the Newton iteration, the iteration converged to a solution, but not one close to the known solution $\sin(x + \pi/6)$. When $U^0(x_i) = \frac{1}{2} - x_i/\pi$ (chosen to satisfy the boundary conditions) was used to start the iteration, convergence to an approximate solution close to the known solution occurred in 5–10 iterations for each value of Δx used. As Δx was halved, the error decreased by a factor of 4.0 each time, confirming the expected $O(\Delta x^2)$ convergence rate.

Table 4.2.1
Errors for Nonlinear Problem

N ($\Delta x = \pi/N$)	Maximum error (after convergence)
25	0.2173E − 02
50	0.5438E − 03
100	0.1357E − 03

Figure 4.2.1 shows the FORTRAN77 program used to solve this problem. The subroutine TRI of Figure 0.4.6, which does partial pivoting, is used to solve the tridiagonal linear system in 4.2.4 for each Newton iteration.

```
      IMPLICIT DOUBLE PRECISION(A-H,O-Z)
C                             N = NUMBER OF GRID POINTS
      PARAMETER (N=100)
      DIMENSION A(N),B(N),C(N),U(0:N),DELTA(N),F(N)
      DATA PI/3.14159265358979312D0/
      DX = PI/N
C                             BOUNDARY VALUES SET
      U(0) =  0.5
      U(N) = -0.5
C                             INITIAL GUESS SATISFIES BOUNDARY CONDITIONS
      DO 10 I=1,N-1
        XI = I*DX
        U(I) = 0.5-XI/PI
   10 CONTINUE
C                             BEGIN NEWTON ITERATION
      DO 50 ITER=1,15
        DO 20 I=1,N-1
C                             CALCULATE F AND THREE DIAGONALS OF JACOBIAN
          F(I) =  (U(I+1)-2*U(I)+U(I-1))/DX**2 -
     &            ((U(I+1)-U(I-1))/(2*DX))**2 - U(I)**2 + U(I) + 1
          A(I) =  1/DX**2 + (U(I+1)-U(I-1))/(2*DX**2)
          B(I) = -2/DX**2 - 2*U(I) + 1
          C(I) =  1/DX**2 - (U(I+1)-U(I-1))/(2*DX**2)
   20   CONTINUE
C                             CALL TRI TO SOLVE TRIDIAGONAL SYSTEM
        CALL TRI(A,B,C,DELTA,F,N-1)
C                             UPDATE SOLUTION AND CALCULATE MAXIMUM ERROR
        ERMAX = 0.0
        DO 30 I=1,N-1
          U(I) = U(I) - DELTA(I)
          XI = I*DX
          ERR = ABS(U(I) - SIN(XI+PI/6.))
          ERMAX = MAX(ERMAX,ERR)
   30   CONTINUE
        PRINT 40, ITER,ERMAX
   40   FORMAT (I5,E15.5)
   50 CONTINUE
      STOP
      END
```

Figure 4.2.1
FORTRAN77 Program to Solve Nonlinear Problem

4.3   A Singular Example

Next we consider the boundary value problem

$$x^2 u_{xx} - 0.11u = 0,$$
$$u(0) = 0, \qquad\qquad\qquad (4.3.1)$$
$$u_x(1) = 1.1,$$

which has the exact solution $u(x) = x^{1.1}$. Using central differences as in 4.1.2
yields ($x_i = i\,\Delta x$, $\Delta x = 1/N$),

$$x_i^2 \frac{U(x_{i+1}) - 2U(x_i) + U(x_{i-1})}{\Delta x^2} - 0.11U(x_i) = 0 \qquad (i = 1, \dots, N),$$

$$U(x_0) = 0, \qquad\qquad\qquad (4.3.2)$$

$$\frac{U(x_{N+1}) - U(x_{N-1})}{2\,\Delta x} = 1.1.$$

The boundary conditions are used to eliminate $U(x_0)$ and $U(x_{N+1})$ from the
first and last equations, respectively, so that there are N unknowns
$U(x_1), \dots, U(x_N)$ and N linear equations:

$$x_1^2 \frac{U(x_2) - 2U(x_1)}{\Delta x^2} - 0.11U(x_1) = 0,$$

$$x_i^2 \frac{U(x_{i+1}) - 2U(x_i) + U(x_{i-1})}{\Delta x^2} - 0.11U(x_i) = 0 \qquad (i = 2, \dots, N-1),$$

$$x_N^2 \frac{2U(x_{N-1}) - 2U(x_N) + 2.2\,\Delta x}{\Delta x^2} - 0.11U(x_N) = 0.$$

Table 4.3.1 shows the maximum error encountered for three values of Δx.
As Δx is halved, the error decreases by a factor of only 2.15 each time. To

Table 4.3.1
Errors for Singular Problem

N ($\Delta x = 1/N$)	Maximum error
25	0.9682E − 03
50	0.4493E − 03
100	0.2091E − 03

calculate the apparent rate of convergence, assume that error $= C \, \Delta x^{\alpha}$, so that

$$0.2091E - 03 = C(0.01)^{\alpha},$$

$$0.4493E - 03 = C(0.02)^{\alpha}.$$

Solving for α gives $2.15 = 2^{\alpha}$, or $\alpha = 1.10$, so that the error appears to be only $O(\Delta x^{1.1})$ rather than the $O(\Delta x^2)$ normally expected when central differences are used.

The reason for this slow convergence is clear when we look at formula 1.2.5b. It shows that our second derivative estimate is guaranteed to be $O(\Delta x^2)$ only if the fourth derivative of the solution is bounded in (x_0, x_N). Since $u(x) = x^{1.1}$, this is not the case, since even the second derivative $u_{xx} = 0.11x^{-0.9}$ is unbounded near $x_0 = 0$. Thus, the truncation error is not guaranteed to be $O(\Delta x^2)$.

Clearly, a uniform mesh spacing, $x_i = i \, \Delta x$, such as has been used for all finite difference formulas thus far, is not appropriate when the solution varies much more rapidly in some portions of the interval than in others, as in this singular problem. Formula 1.2.5b suggests that Δx should be made smallest where the fourth derivative of the solution is largest, so that the decrease in Δx will offset the increase in u_{xxxx}. However, while accurate finite difference formulas on nonuniform grids do exist, the nonuniformity of the grid penalizes any straightforward attempt to approximate the second derivative of u with either a loss of accuracy or the inclusion of extra unknowns per equation (Problem 5).

One of the easiest ways to derive accurate finite difference formulas on variable grids is (ironically) to use the finite element method (see Problem 2 of Chapter 5). We will therefore defer treatment of nonuniform grids to Chapter 5. In fact, the singular problem 4.3.1 is solved accurately in Section 5.4 by using a finite element method with a nonuniform grid.

It should be mentioned that the tridiagonal linear system solver that does no pivoting (Figure 0.4.4) can be safely used on the system 4.3.2, since it is diagonal dominant.

4.4 Shooting Methods

For boundary value problems in one space variable, there is an alternative class of methods, called "shooting" methods—so called for reasons that will soon become evident. To illustrate the use of these methods, consider the following second-order, nonlinear, two-point boundary value problem, which is a generalization of 4.1.1. The use of shooting methods on more general problems is illustrated by Problem 7.

$$u'' = f(x, u, u'),$$
$$u(0) = g_0,$$
$$u(1) = g_1.$$
(4.4.1)

A shooting method works with the *initial value* problem

$$u'' = f(x, u, u'),$$
$$u(0) = g_0,$$
$$u'(0) = \beta,$$
(4.4.2)

which is usually treated as a system of two first-order equations:

$$u' \equiv v,$$
$$v' = f(x, u, v),$$
$$u(0) = g_0,$$
$$v(0) = \beta.$$
(4.4.3)

This initial value problem is solved for various values of β, with the hope of finding a value of β that gives a solution satisfying $u(1) = g_1$. This is somewhat like shooting a cannon, stationed at the point $(0, g_0)$, trying different shooting angles until the target at the point $(1, g_1)$ is hit (see Figure 4.4.1). Of course, it would be foolish to do this simply by trial and error—we must learn from our misses. In fact, we are really trying to find a root of the nonlinear function

$$R(\beta) \equiv u_\beta(1) - g_1,$$

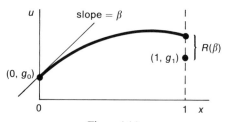

Figure 4.4.1

where $u_\beta(x)$ represents the solution of the initial value problem 4.4.2. An appropriate method for finding a root of $R(\beta)$ might be the secant iteration

$$\beta_{n+1} = \beta_n - \frac{R(\beta_n)(\beta_n - \beta_{n-1})}{R(\beta_n) - R(\beta_{n-1})}, \qquad (4.4.4)$$

which converges rapidly near a root and requires no derivatives of $R(\beta)$.

If the differential equation is linear, then $R(\beta)$ will be a linear function of β, since $u_\beta(x) = u_0(x) + \beta w(x)$ and thus $R(\beta) = u_0(1) + \beta w(1) - g_1$, where $u_0(x)$ is the solution of 4.4.2 with $\beta = 0$ and $w(x)$ is the solution of the corresponding homogeneous ordinary differential equation satisfying $w(0) = 0$, $w'(0) = 1$. Therefore, only two "shots," with $\beta = \beta_1$ and $\beta = \beta_2$, are necessary, since β_3, calculated according to the secant method 4.4.4, will be exact, ignoring roundoff error and errors committed by the initial value solver.

This shooting method was applied to solve the linear problem

$$u'' = 400u - 401 \sin(x),$$

$$u(0) = 0, \qquad (4.4.5)$$

$$u(1) = \sin(1),$$

which has the exact solution $u(x) = \sin(x)$. The corresponding initial value problem, reduced to a first-order system as in 4.4.3, is

$$
\begin{aligned}
u' &= v, & u(0) &= 0, \\
v' &= 400u - 401 \sin(x), & v(0) &= \beta.
\end{aligned}
\qquad (4.4.6)
$$

The initial value solver IVPRK, taken from the IMSL Math Library, was used to solve this system for each value of β. IVPRK uses a Runge–Kutta type method (see Section 1.6), and automatically varies the stepsize to control the error.

The initial value problem 4.4.6 was solved twice, with $\beta_1 = 3$ and $\beta_2 = 2$. The secant formula 4.4.4 then gave $\beta_3 = 1.0$ correct to four decimal places (since $u(x) = \sin(x)$, $u'(0) = 1$). Single precision on an IBM 4341 computer was used. When the problem 4.4.6 was re-solved with β set exactly equal to 1, however, the initial value solver missed the target point $(1, \sin(1))$ by a factor of 100, even with the error tolerance for IVPRK set as small as it could handle! Shooting in the reverse direction, that is, trying to determine β such that the solution of the differential equation with $u(1) = \sin(1)$, $u'(1) = \beta$ satisfies $u(0) = 0$, was equally unsuccessful.

In order to understand the problem, observe that the analytic solution to 4.4.6 has

$$u_\beta(x) = \frac{(\beta - 1)[e^{20x} - e^{-20x}]}{40} + \sin(x),$$

so that

$$R(\beta) = u_\beta(1) - \sin(1) = \frac{(\beta - 1)[e^{20} - e^{-20}]}{40} = 1.2 \times 10^7(\beta - 1).$$

To say that the function $R(\beta)$ is steep near the root $\beta = 1$ is an understatement. A very small inaccuracy in the angle β causes $u_\beta(1)$ to miss the target by a large margin. Even when β is set exactly to 1, small and unavoidable errors introduced by the initial value solver are magnified greatly at $x = 1$. Thus, even though the shooting method allowed the accurate determination of β, this parameter is numerically useless in predicting the form of the solution between 0 and 1.

This sort of difficulty is not at all uncommon, and it illustrates why "simple" shooting methods are not recommended for general boundary value problems. The difficulty is clearly the presence of the e^{20x} term in the homogeneous solution. (Reverse shooting has trouble with the e^{-20x} term.) These terms did not, however, create any difficulties for the finite difference method 4.1.2, when it was used to solve 4.4.5.

A technique that is often useful when simple shooting fails is called "multiple" or "parallel" shooting. The interval (say (0, 1)) is subdivided into a (moderate) number of subintervals using the knots $0 = x_0 < x_1 < \cdots < x_M = 1$. Assuming a boundary value problem of the form 4.4.1, the multiple shooting method works on the initial value problems

$$\text{in } x_0 \leq x < x_1: \qquad u' = v, \qquad\qquad u(x_0) = g_0,$$
$$v' = f(x, u, v), \qquad v(x_0) = \beta_0,$$
$$\text{in } x_k \leq x < x_{k+1}: \qquad u' = v, \qquad\qquad u(x_k) = \alpha_k,$$
$$(k = 1, \ldots, M - 1) \qquad v' = f(x, u, v), \qquad v(x_k) = \beta_k.$$

The numbers $\beta_0, \alpha_1, \beta_1, \ldots, \alpha_{M-1}, \beta_{M-1}$ are called the shooting parameters. For any given choice of these parameters, the initial value problem is solved from 0 to 1. At each interior knot the residuals

$$R_k = u(x_k-) - \alpha_k \qquad (k = 1, \ldots, M - 1),$$
$$S_k = v(x_k-) - \beta_k,$$

are calculated, which measure the discontinuities in u and v, since $u(x_{k+}) = \alpha_k$, $v(x_{k+}) = \beta_k$. At x_M $(=1)$ the residual

$$R_M = u(x_{M-}) - g_1$$

is calculated, which measures the amount by which the second boundary condition fails to be satisfied.

If the $2M - 1$ shooting parameters are somehow chosen so that the $2M - 1$ residuals $R_1, S_1, \ldots, R_{M-1}, S_{M-1}, R_M$ are zero, then functions u, v have been found which are continuous and satisfy the differential equation and both boundary conditions. Therefore, we need to find the solution to a system of $2M - 1$ equations with $2M - 1$ unknown parameters.

IMSL routine BVPMS [Sewell, 1982] uses a multiple shooting method, with IVPRK used to solve the initial value problems, to solve general systems of two-point boundary value problems with general boundary conditions. It uses Newton's method to solve for the shooting parameters; the Jacobian matrix elements are found by solving the system of differential equations that results when the original differential equations are differentiated with respect to the shooting parameters. The resulting Jacobian matrix is banded. This subroutine was used to test the multiple shooting method applied to the problem 4.4.5.

With $x_0 = 0$, $x_1 = 1$ (simple shooting), BVPMS gave terrible results, with the proper error message. For example, it returned $v(1) = 238$ when the true value is $\cos(1)$. With four intermediate shooting points $(x_k = 0.2k, k = 0, \ldots, 5)$ BVPMS gave good results, with $v(1)$ estimated correctly to five significant digits. Single precision was used, as in the simple shooting experiments.

Intuitively, the reason that multiple shooting works better than simple shooting, in this case, is that exponentially growing components of the solution have less "time" to grow large before they are reinitialized. For example, the component e^{20x} in the general solution to 4.4.5 can only increase in size by a factor of $e^4 \approx 55$ on a subinterval of length 0.2, as compared to $e^{20} \approx 4.9 \times 10^8$ on the entire interval $[0, 1]$.

The principle advantage of shooting methods over finite difference or finite element methods is that they transfer the problems of error control and stepsize variation to the initial value solver, relieving the boundary value code developer of these difficult and critical tasks. Initial value solvers of great sophistication and reliability are widely available (the NAG, IMSL, Harwell, and SLATEC libraries all have at least one such code). Of course, there is no reason why a finite difference or finite element boundary value solver cannot take care of the error control and stepsize variation itself, and in fact such codes do exist. For example, both the IMSL and NAG libraries contain versions of Pereyra and Lentini's PASVAR program [Pereyra, 1978], a sophisticated finite difference code that adaptively determines the density and distribution of the grid points and uses "deferred corrections" (an idea remotely related to Romberg integration) to obtain high-order finite difference approximations.

4.5   Multi-Dimensional Problems

The finite difference methods discussed in Section 4.1 generalize in a straightforward way to multi-dimensional problems. For example, the two-dimensional steady-state diffusion problem, with positive, constant, diffusion coefficient (D),

$$0 = Du_{xx} + Du_{yy} + f(x, y) \qquad \text{in } (0, 1) \times (0, 1),$$
$$\text{with } u = g(x, y) \qquad\qquad \text{on the boundary,}$$

can be discretized in a manner similar to formula 2.4.2:

$$-D\,\frac{U(x_{i+1}, y_j) - 2U(x_i, y_j) + U(x_{i-1}, y_j)}{\Delta x^2}$$

$$-D\,\frac{U(x_i, y_{j+1}) - 2U(x_i, y_j) + U(x_i, y_{j-1})}{\Delta y^2} = f(x_i, y_j). \quad (4.5.1)$$

Here

$$x_i = i\Delta x \qquad \Delta x = 1/N,$$
$$y_j = j\Delta y \qquad \Delta y = 1/M,$$

and the values (circled in Figure 4.5.1) of $U(x_i, y_j)$ are known on the boundary.

Thus there are a total of $(N - 1)(M - 1)$ interior points and an equal number of unknowns and equations, since one equation is centered on each unknown.

It is interesting to note that for Laplace's equation ($f = 0$), 4.5.1 shows that $U(x_i, y_j)$ is equal to the average (weighted average, if $\Delta x \neq \Delta y$) of the values of U at its four neighboring points. This mirrors a property of the exact solution (see Section 4.0) and shows also that U, like u, cannot attain a local extremum in the interior of the region.

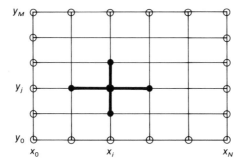

Figure 4.5.1

The system of linear equations 4.5.1 has exactly the same nonzero structure as the system 2.4.4 which has to be solved at each time step when an implicit method is used to solve the time-dependent diffusion equation. If $\Delta x = \Delta y = 0.2$ and the unknowns are ordered in "typewriter" order, the resulting system is similar to that of Figure 2.4.2, and has the form shown in Figure 4.5.2.

If the three-dimensional partial differential equation

$$0 = Du_{xx} + Du_{yy} + Du_{zz} + f(x, y, z) \qquad \text{in } (0, 1) \times (0, 1) \times (0, 1)$$

is approximated by using a second-order centered finite difference formula, with $\Delta x = \Delta y = \Delta z = 0.25$, a linear system results whose coefficient matrix is shown in Figure 4.5.3. It is assumed that the unknowns are ordered in typewriter order on each $z = $ constant plane, one plane at a time.

When $N = 1/\Delta x = 1/\Delta y$, the two-dimensional linear system 4.5.1 has rank $(N - 1)^2$ and half-bandwidth (see Section 0.4) $L = N - 1$, so the total work to solve it by using a band solver is the rank times L^2, which is $O(N^4)$. When $N = 1/\Delta x = 1/\Delta y = 1/\Delta z$, the three-dimensional linear system has rank $(N - 1)^3$ and half-bandwidth of $(N - 1)^2$, so the total work done by the band solver is $O(N^7)$. (For nonlinear problems, these estimates represent the amount of work to solve the required linear system each Newton iteration.) Comparing these to the $O(N)$ work required to solve the one-dimensional equations, we see that efficiency in solving the multi-dimensional linear systems is much more crucial than for the (typically tridiagonal) linear systems arising from the one-dimensional problems.

Fortunately, there are a number of iterative methods that (in addition to the very complicated sparse direct solvers mentioned in Section 0.5) can solve systems such as 4.5.1 more efficiently than a band solver, and some are also

$$
\begin{bmatrix}
4 & -1 & 0 & 0 & -1 & 0 & 0 & 0 & 0 & 0 & 0 & 0 & 0 & 0 & 0 & 0 \\
-1 & 4 & -1 & 0 & 0 & -1 & 0 & 0 & 0 & 0 & 0 & 0 & 0 & 0 & 0 & 0 \\
0 & -1 & 4 & -1 & 0 & 0 & -1 & 0 & 0 & 0 & 0 & 0 & 0 & 0 & 0 & 0 \\
0 & 0 & -1 & 4 & 0 & 0 & 0 & -1 & 0 & 0 & 0 & 0 & 0 & 0 & 0 & 0 \\
-1 & 0 & 0 & 0 & 4 & -1 & 0 & 0 & -1 & 0 & 0 & 0 & 0 & 0 & 0 & 0 \\
0 & -1 & 0 & 0 & -1 & 4 & -1 & 0 & 0 & -1 & 0 & 0 & 0 & 0 & 0 & 0 \\
0 & 0 & -1 & 0 & 0 & -1 & 4 & -1 & 0 & 0 & -1 & 0 & 0 & 0 & 0 & 0 \\
0 & 0 & 0 & -1 & 0 & 0 & -1 & 4 & 0 & 0 & 0 & -1 & 0 & 0 & 0 & 0 \\
0 & 0 & 0 & 0 & -1 & 0 & 0 & 0 & 4 & -1 & 0 & 0 & -1 & 0 & 0 & 0 \\
0 & 0 & 0 & 0 & 0 & -1 & 0 & 0 & -1 & 4 & -1 & 0 & 0 & -1 & 0 & 0 \\
0 & 0 & 0 & 0 & 0 & 0 & -1 & 0 & 0 & -1 & 4 & -1 & 0 & 0 & -1 & 0 \\
0 & 0 & 0 & 0 & 0 & 0 & 0 & -1 & 0 & 0 & -1 & 4 & 0 & 0 & 0 & -1 \\
0 & 0 & 0 & 0 & 0 & 0 & 0 & 0 & -1 & 0 & 0 & 0 & 4 & -1 & 0 & 0 \\
0 & 0 & 0 & 0 & 0 & 0 & 0 & 0 & 0 & -1 & 0 & 0 & -1 & 4 & -1 & 0 \\
0 & 0 & 0 & 0 & 0 & 0 & 0 & 0 & 0 & 0 & -1 & 0 & 0 & -1 & 4 & -1 \\
0 & 0 & 0 & 0 & 0 & 0 & 0 & 0 & 0 & 0 & 0 & -1 & 0 & 0 & -1 & 4 \\
\end{bmatrix}
\begin{bmatrix}
U11 \\ U21 \\ U31 \\ U41 \\ U12 \\ U22 \\ U32 \\ U42 \\ U13 \\ U23 \\ U33 \\ U43 \\ U14 \\ U24 \\ U34 \\ U44
\end{bmatrix}
=
\begin{bmatrix}
\Delta x^2 F11/D + U01 + U10 \\
\Delta x^2 F21/D + U20 \\
\Delta x^2 F31/D + U30 \\
\Delta x^2 F41/D + U51 + U40 \\
\Delta x^2 F12/D + U02 \\
\Delta x^2 F22/D \\
\Delta x^2 F32/D \\
\Delta x^2 F42/D + U52 \\
\Delta x^2 F13/D + U03 \\
\Delta x^2 F23/D \\
\Delta x^2 F33/D \\
\Delta x^2 F43/D + U53 \\
\Delta x^2 F14/D + U04 + U15 \\
\Delta x^2 F24/D + U25 \\
\Delta x^2 F34/D + U35 \\
\Delta x^2 F44/D + U54 + U45
\end{bmatrix}
$$

where $Uij = U(x_i, y_j)$ $Fij = f(x_i, y_j)$

Figure 4.5.2
Linear System for 2-D Finite Difference Method

Figure 4.5.3
Coefficient Matrix for 3-D Finite Difference Method

even easier to program. Two of the more popular of these iterative methods will be discussed in the following sections.

Since many of these iterative methods (and even the direct solvers) have improved properties when the linear system has a positive-definite (or negative-definite) matrix, it should be mentioned here that the linear system 4.5.1 is positive definite. To see this, note that the matrix (call it A) of the linear system 4.5.1 is symmetric and weakly diagonal dominant, with positive diagonal entries. (When $\Delta x = \Delta y$, A takes the form shown in Figure 4.5.2.) This guarantees that the eigenvalues are real and nonnegative, for if $\lambda < 0$, the matrix $(A - \lambda I)$ is clearly (strongly) diagonal dominant, and therefore nonsingular by Theorem 0.2.1; thus $\lambda < 0$ cannot be an eigenvalue. Using an argument similar to that used to show that 4.1.2 represents a nonsingular linear system (cf. Problem 8), it can be shown that A is nonsingular, and thus $\lambda = 0$ cannot be an eigenvalue either. Therefore A must be positive definite.

As long as all boundaries are of the form $\{x, y, \text{or } z\} = $ constant, finite difference methods are fairly easy to apply to multi-dimensional regions. When the boundary is curved, however, we prefer to use finite element methods (see Chapter 5), which handle such regions more naturally.

4.6   Successive Over-Relaxation

One of the most popular simple iterative methods for solving the linear systems generated by boundary value applications is called Successive Over-Relaxation (SOR). We begin by looking at a generalization of this method, called Nonlinear Over-Relaxation (NLOR), which can be used to solve nonlinear systems of the form

$$f_k(u_1, \ldots, u_M) = 0, \qquad k = 1, \ldots, M. \tag{4.6.1}$$

The NLOR iteration for this system is

$$u_k^{n+1} = u_k^n - \omega \frac{f_k(u_1^{n+1}, \ldots, u_{k-1}^{n+1}, u_k^n, \ldots, u_M^n)}{\dfrac{\partial f_k}{\partial u_k}(u_1^{n+1}, \ldots, u_{k-1}^{n+1}, u_k^n, \ldots, u_M^n)} \qquad k = 1, \ldots, M. \tag{4.6.2}$$

Clearly, if 4.6.2 converges to a vector $u_1, \ldots, u_M$, then $f_k(u_1, \ldots, u_M) = 0$, so that the limit is, in fact, a solution.

This formula is similar to the Newton method iteration with the Jacobian replaced by its diagonal part, except that there is an extra parameter ω

involved (whose value will be discussed presently), and the functions are not simply evaluated at $\mathbf{u}^n$. Thus it is not surprising that our convergence criteria will require that the diagonal part of this Jacobian be large, in some sense, relative to the rest of the matrix. Clearly NLOR is not designed for general nonlinear problems, since there is not even any guarantee that $\partial f_k / \partial u_k$ will not be identically zero for some values of k.

Contrary to the initial impression, 4.6.2 is not implicit, since by the time u_k^{n+1} is calculated, $u_1^{n+1}, \ldots, u_{k-1}^{n+1}$ are known quantities. The functions f_k and $\partial f_k / \partial u_k$ are evaluated using the *latest available* values of the components of $\mathbf{u}$.

If the system 4.6.1 is linear, $A\mathbf{u} - \mathbf{b} = 0$; that is, if

$$f_k = \sum_{j=1}^{M} a_{kj} u_j - b_k = 0, \qquad k = 1, \ldots, M,$$

then the iteration, now called SOR, takes the form

$$u_k^{n+1} = u_k^n - \omega \frac{\left[\sum_{j=1}^{k-1} a_{kj} u_j^{n+1} + a_{kk} u_k^n + \sum_{j=k+1}^{M} a_{kj} u_j^n - b_k \right]}{a_{kk}}. \tag{4.6.3}$$

If $\mathbf{u}$ is the exact solution of $A\mathbf{u} - \mathbf{b} = 0$, then it is easy to see that 4.6.3 still holds with u_k^{n+1} and u_k^n replaced by u_k, since the quantity in brackets vanishes. Thus if $e_k^n \equiv u_k^n - u_k$, subtracting the equation satisfied by $\mathbf{u}$ from 4.6.3 yields an equation for the error:

$$e_k^{n+1} = e_k^n - \omega \frac{\left[\sum_{j=1}^{k-1} a_{kj} e_j^{n+1} + a_{kk} e_k^n + \sum_{j=k+1}^{M} a_{kj} e_j^n \right]}{a_{kk}}. \tag{4.6.4}$$

If the matrix A is broken down into its (strictly) lower triangular part (L), diagonal part (D), and upper triangular part (U), $A = L + D + U$, then 4.6.4 can be written in the matrix–vector form

$$\mathbf{e}^{n+1} = \mathbf{e}^n - \omega D^{-1}\{L\mathbf{e}^{n+1} + D\mathbf{e}^n + U\mathbf{e}^n\}$$

or

$$\mathbf{e}^{n+1} = (D + \omega L)^{-1}(D - \omega D - \omega U)\mathbf{e}^n = \cdots$$

$$= [(D + \omega L)^{-1}(D - \omega D - \omega U)]^{n+1} \mathbf{e}^0$$

Since it is known that the powers of a matrix converge to zero if and only if all the matrix eigenvalues are less than 1 in absolute value [Horn and Johnson,

1985, Theorem 5.6.12], we see that the convergence of the successive over-relaxation method depends on the eigenvalues of the matrix $(D + \omega L)^{-1}(D - \omega D - \omega U)$. The following result states a condition that guarantees convergence.

Theorem 4.6.1 *If* $0 < \omega < 2$ *and* $A = L + D + U$ *is positive definite, all eigenvalues of* $(D + \omega L)^{-1}(D - \omega D - \omega U)$ *are less than 1 in absolute value, and the successive over-relaxation method applied to* $A\mathbf{u} - \mathbf{b} = 0$ *converges.*

Proof First note that if A is positive-definite, and thus symmetric, $U = L^T$. Now suppose that λ, $\mathbf{z}$ is a (possibly complex) eigenvalue–eigenvector pair, that is

$$\{(1 - \omega)D - \omega L^T\}\mathbf{z} = \lambda(D + \omega L)\mathbf{z}. \tag{4.6.5}$$

Taking the scalar product of both sides with $\bar{\mathbf{z}}$ (the conjugate of $\mathbf{z}$),

$$(1 - \omega)\bar{\mathbf{z}}^T D\mathbf{z} - \omega\bar{\mathbf{z}}^T L^T\mathbf{z} = \lambda\bar{\mathbf{z}}^T D\mathbf{z} + \lambda\omega\bar{\mathbf{z}}^T L\mathbf{z}$$

or

$$(1 - \omega)d - \omega\bar{l} = \lambda d + \lambda\omega l, \tag{4.6.6}$$

where $d \equiv \bar{\mathbf{z}}^T D\mathbf{z}$ and $l \equiv \bar{\mathbf{z}}^T L\mathbf{z}$. Also, since $A = L + D + L^T$,

$$\bar{\mathbf{z}}^T A\mathbf{z} = \bar{\mathbf{z}}^T L\mathbf{z} + \bar{\mathbf{z}}^T D\mathbf{z} + \bar{\mathbf{z}}^T L^T\mathbf{z}$$

or

$$a = l + d + \bar{l}, \tag{4.6.7}$$

where d and l are as defined above, and $a \equiv \bar{\mathbf{z}}^T A\mathbf{z}$.

Since A is positive-definite, its diagonal, D, has positive elements. Thus $d = \bar{\mathbf{z}}^T D\mathbf{z} = \sum D_{ii}|z_i|^2$, which is real and positive. Also, the positive-definiteness of A implies that $A = S^T E S$, where E is a positive diagonal matrix containing the eigenvalues of A. Thus

$$a = \bar{\mathbf{z}}^T S^T E S\mathbf{z} = \bar{\mathbf{y}}^T E\mathbf{y} = \sum E_{ii}|y_i|^2,$$

where $\mathbf{y} \equiv S\mathbf{z}$, so that a is real and positive likewise.

Now l can be eliminated between 4.6.6 and 4.6.7, taking into account the fact that a and d are real. After considerable effort, we get

$$1 - |\lambda|^2 = \frac{d(2 - \omega)|1 - \lambda|^2}{a\omega}. \tag{4.6.8}$$

Since $0 < \omega < 2$ and a and d are positive, the right-hand side is nonnegative. Further, λ cannot equal 1, for if $\lambda = 1$ were an eigenvalue of 4.6.5, we would have

$$\mathbf{0} = \omega(L + D + L^T)\mathbf{z} = \omega A \mathbf{z}$$

for some nonzero $\mathbf{z}$ (the eigenvector), contradicting the assumption that A is positive-definite, and thus nonsingular. Therefore, the right-hand side of 4.6.8 is positive, and so finally

$$|\lambda|^2 < 1$$

for any eigenvalue of 4.6.5.   ■

Another condition on A that guarantees convergence, at least for certain values of the parameter ω, is diagonal dominance, as the following result shows.

Theorem 4.6.2 *If $0 < \omega \le 1$ and A is diagonal-dominant, the successive over-relaxation method applied to $A\mathbf{u} - \mathbf{b} = \mathbf{0}$ converges.*

Proof Let us define

$$r_k \equiv \sum_{j=1}^{k-1} \frac{|a_{kj}|}{|a_{kk}|}, \qquad s_k \equiv \sum_{j=k+1}^{M} \frac{|a_{kj}|}{|a_{kk}|},$$

Then, because A is diagonal-dominant, $r_k + s_k \le \rho < 1$, for all k. Thus from 4.6.4, we have (since $1 - \omega \ge 0$)

$$|e_k^{n+1}| \le \omega r_k \|e^{n+1}\|_\infty + (1 - \omega + \omega s_k)\|e^n\|_\infty.$$

Now let l be an index for which $|e_l^{n+1}|$ attains its maximum value $\|e^{n+1}\|_\infty$. Then,

$$\|e^{n+1}\|_\infty \le \omega r_l \|e^{n+1}\|_\infty + (1 - \omega + \omega s_l)\|e^n\|_\infty$$

or, since $1 - \omega r_l$ is positive,

$$\|e^{n+1}\|_\infty \le \frac{(1 - \omega + \omega s_l)}{(1 - \omega r_l)}\|e^n\|_\infty.$$

The error reduction factor is bounded by

$$\frac{(1 - \omega + \omega s_l)}{(1 - \omega r_l)} \leq \frac{1 - \omega + \omega(\rho - r_l)}{(1 - \omega r_l)}$$

$$= 1 - \frac{\omega(1 - \rho)}{(1 - \omega r_l)} \leq 1 - \omega(1 - \rho).$$

Since this is less than 1, under our assumptions, the error is converging to zero, as claimed. ∎

Most of the nonsymmetric, diagonal-dominant linear systems encountered solving partial differential equations are so nearly positive-definite that, despite Theorem 4.6.2, SOR will converge for some values of ω greater than 1. In fact, the best choice of the parameter ω will nearly always be greater than 1. When $\omega = 1$, the successive over-relaxation iteration is called the Gauss–Seidel method.

4.7   Successive Over-Relaxation Examples

Now let us consider the problem

$$-u_{xx} - u_{yy} + au - xy(ay^2 - 6) = 0 \qquad \text{in } (0, 1) \times (0, 1),$$

with

$$u(x, 0) = 0,$$

$$u(x, 1) = x,$$

$$u(0, y) = 0,$$

$$u(1, y) = y^3,$$

where a is a constant to be assigned a value later. The exact solution is $u = xy^3$. This partial differential equation is approximated by the finite difference formula

$$-\left[\frac{U(x_{i+1}, y_j) - 2U(x_i, y_j) + U(x_{i-1}, y_j)}{\Delta x^2}\right]$$

$$-\left[\frac{U(x_i, y_{j+1}) - 2U(x_i, y_j) + U(x_i, y_{j-1})}{\Delta y^2}\right]$$

$$+ aU(x_i, y_j) - x_i y_j(ay_j^2 - 6) = 0 \qquad \text{(for } i = 1, \ldots, N - 1, j = 1, \ldots, M - 1),$$

$$(4.7.1)$$

with

$$U(x_i, y_0) = 0,$$

$$U(x_i, y_M) = x_i,$$

$$U(x_0, y_j) = 0,$$

$$U(x_N, y_j) = y_j^3,$$

where $\Delta x = 1/N$, $\Delta y = 1/M$, $x_i = i\,\Delta x$, $y_j = j\,\Delta y$.

Recalling that the truncation error is the amount by which the exact solution of the partial differential equation fails to satisfy the discrete equations, we calculate, using a bivariate Taylor series expansion (cf. 1.2.5b):

$$T_{ij} = -\left[u_{xx}(x_i, y_j) + \frac{\Delta x^2}{12} u_{xxxx}(\alpha_{ij}, y_j)\right]$$

$$- \left[u_{yy}(x_i, y_j) + \frac{\Delta y^2}{12} u_{yyyy}(x_i, \beta_{ij})\right] + au(x_i, y_j) - x_i y_j(a y_j^2 - 6)$$

$$= -\frac{\Delta x^2}{12} u_{xxxx}(\alpha_{ij}, y_j) - \frac{\Delta y^2}{12} u_{yyyy}(x_i, \beta_{ij}).$$

Since $u = xy^3$, u_{xxxx} and u_{yyyy} are identically zero, so the truncation error is identically zero. Thus U should be exactly equal to u at the nodes (x_i, y_j), no matter how large Δx and Δy may be. This is by design, so that when the SOR iterative method is used to solve 4.7.1, all errors can be attributed to incomplete convergence of the iterative method.

The structure of the matrix of the linear system 4.7.1 is similar to that of Figure 4.5.2. If the equations are ordered in the same way as the unknowns on which they are centered (we will always do this), this matrix will be symmetric, even if $\Delta x \neq \Delta y$, and even if the region is not rectangular. To see this, let K and L be the numbers of any two interior nodes. If nodes K and L are "vertical neighbors," that is, they have the same x coordinate and y coordinates differing by Δy, then $A_{KL} = -1/\Delta y^2 = A_{LK}$, from 4.7.1. Similarly, if they are "horizontal neighbors," $A_{KL} = -1/\Delta x^2 = A_{LK}$. If they are neither vertical nor horizontal neighbors (and $K \neq L$), $A_{KL} = A_{LK} = 0$. If first-order derivatives are added to the partial differential equation, the discrete matrix is no longer symmetric.

To determine if our matrix is positive-definite, we calculate its eigenvalues directly. The eigenvectors of this $(N-1)(M-1)$ by $(N-1)(M-1)$ matrix are $\mathbf{U}_{kl}$, $k = 1, \ldots, N-1$, $l = 1, \ldots, M-1$, where $\mathbf{U}_{kl}$ has elements

$$U_{kl}(x_i, y_j) = \sin(k\pi x_i)\sin(l\pi y_j).$$

Since 4.7.1 is a linear system of the form $A\mathbf{U} + \mathbf{b} = \mathbf{0}$, the homogeneous portion of 4.7.1 represents $A\mathbf{U}$. To calculate the i,jth component of $A\mathbf{U}_{kl}$ and to verify that $\mathbf{U}_{kl}$ is an eigenvector, simply replace $U(x_i, y_j)$ in the homogeneous part of 4.7.1 by the i,jth component of $\mathbf{U}_{kl}$. After using some trigonometric identities,

$$
-\left[\frac{U_{kl}(x_{i+1}, y_j) - 2U_{kl}(x_i, y_j) + U_{kl}(x_{i-1}, y_j)}{\Delta x^2}\right]
$$

$$
-\left[\frac{U_{kl}(x_i, y_{j+1}) - 2U_{kl}(x_i, y_j) + U_{kl}(x_i, y_{j-1})}{\Delta y^2}\right] + aU_{kl}(x_i, y_j)
$$

$$
= \left[\frac{4 \sin^2(\tfrac{1}{2}k\pi\,\Delta x)}{\Delta x^2} + \frac{4 \sin^2(\tfrac{1}{2}l\pi\,\Delta y)}{\Delta y^2} + a\right]U_{kl}(x_i, y_j).
$$

Since it also satisfies the homogeneous boundary conditions, so that the cases $i = 1$, $i = N - 1$, $j = 1$, $j = M - 1$ need not be treated separately, $\mathbf{U}_{kl}$ is an eigenvector as claimed, and the corresponding eigenvalue is

$$
\lambda_{kl} = \frac{4}{\Delta x^2} \sin^2(\tfrac{1}{2}k\pi\,\Delta x) + \frac{4}{\Delta y^2} \sin^2(\tfrac{1}{2}l\pi\,\Delta y) + a. \qquad (4.7.2)
$$

Assuming Δx and Δy are very small, the (algebraically) smallest eigenvalue λ_{11} is approximately

$$
\lambda_{11} = \frac{4}{\Delta x^2} \sin^2(\tfrac{1}{2}\pi\,\Delta x) + \frac{4}{\Delta y^2} \sin^2(\tfrac{1}{2}\pi\,\Delta y) + a
$$

$$
\approx 2\pi^2 + a. \qquad (4.7.3)
$$

Thus, as long as a is greater than approximately $-2\pi^2 = -19.74$, the matrix for 4.7.1 is positive-definite and, by Theorem 4.6.1, successive over-relaxation is guaranteed to converge when applied to 4.7.1, provided $0 < \omega < 2$. Of course, as yet we have no evidence that it will converge fast enough to be useful.

The linear system 4.7.1 was solved using SOR (formula 4.6.3), with various values for a, ω and $\Delta x = \Delta y$. The FORTRAN77 program for one of these experiments is shown in Figure 4.7.1. Notice the extraordinary simplicity of the program. The known boundary values of U and the unknown interior values (cf. Figure 4.5.1) are stored conveniently together in the same array. Notice also that the requirement in 4.6.3 that the latest calculated values of the unknowns be used in evaluating f_k and $\partial f_k / \partial u_k$, is handled effortlessly in the program (DO loop 15). Zero starting values are used.

```
      IMPLICIT DOUBLE PRECISION(A-H,O-Z)
C                               N = NUMBER OF POINTS IN X,Y DIRECTIONS
C                               A = PDE PARAMETER
      PARAMETER (N=40, A=40.0)
      DIMENSION U(0:N,0:N),X(0:N),Y(0:N)
      DX = 1.D0/N
      DY = DX
C                               TRY VARIOUS VALUES OF THE PARAMETER OMEGA
      DO 35 IW=0,19
        W = 1.0 + IW*0.05
C                               SET UNKNOWNS TO ZERO STARTING VALUES
        DO 5 I=1,N-1
        DO 5 J=1,N-1
    5   U(I,J) = 0.0
C                               SET VALUES OF U ON BOUNDARY (THESE WILL NEVER
C                               CHANGE)
        DO 10 L=0,N
          X(L) = L*DX
          Y(L) = L*DY
          U(L,0) = 0.0
          U(L,N) = X(L)
          U(0,L) = 0.0
          U(N,L) = Y(L)**3
   10   CONTINUE
C                               BEGIN SOR ITERATION
        DO 20 ITER=1,10000
          ERMAX = 0.0
C                               UPDATE UNKNOWNS USING SOR FORMULA
          DO 15 I=1,N-1
          DO 15 J=1,N-1
            U(I,J) = U(I,J) - W*( (-U(I+1,J)+2*U(I,J)-U(I-1,J))/DX**2
     &                           +(-U(I,J+1)+2*U(I,J)-U(I,J-1))/DY**2
     &         + A*U(I,J) - X(I)*Y(J)*(A*Y(J)**2-6) ) / (4/DX**2+A)
            ERR = ABS(U(I,J) - X(I)*Y(J)**3)
            ERMAX = MAX(ERMAX,ERR)
   15     CONTINUE
          MAXITR = ITER
C                             STOP IF ERROR .LT. 10**(-6)
          IF (ERMAX.LT.1.D-6) GO TO 25
   20   CONTINUE
   25   PRINT 30, W,MAXITR
   30   FORMAT (5X,F8.2,I10)
   35 CONTINUE
      STOP
      END
```

Figure 4.7.1
FORTRAN77 Program Implementing SOR Iteration

The order in which the unknowns are updated in the SOR iteration affects the calculations, according to 4.6.3. However, simultaneously reordering the unknowns and their associated equations transforms the coefficient matrix A into $P^T A P$, where P is a unitary ($P^T = P^{-1}$) permutation matrix. Thus if A is positive-definite, the reordered matrix is still positive-definite, with the same eigenvalues, and the updating order should not greatly affect the convergence rate.

The number of SOR iterations required to reduce the maximum error to less than 10^{-6} for each set of parameters (a, ω, Δx) is shown in Table 4.7.1. The exact solution of 4.7.1 is known, since $U(x_i, y_j) = u(x_i, y_j) = x_i y_j^3$.

For $a = 0$ and $a = 40$, the number of iterations required appears to be proportional to N ($= 1/\Delta x$), when ω is given its "optimal" value each time (marked by asterisks). As N is doubled, the number of iterations also doubles, approximately, as seen in Table 4.7.2, which summarizes the results of Table 4.7.1.

If the number of iterations is really $O(N)$, as observed, then the total work to solve the linear system by using successive over-relaxation is $O(N^3)$, because there are $O(N^2)$ operations per iteration (the number of unknowns is $O(N^2)$). This compares favorably with the $O(N^4)$ work that would be required by a band solver on this problem.

Even if ω is simply chosen equal to 1 (the Gauss–Seidel method), Table 4.7.2 suggests that the number of iterations is $O(N^2)$, since this number quadruples as N is doubled. This makes the total work $O(N^4)$.

In fact, it can be shown that, for this problem, the optimal choice for ω is [Forsythe and Wasow, 1960, Section 22.1]:

$$\omega_{opt} = \frac{2}{1 + (1 - \lambda_J^2)^{1/2}} \tag{4.7.4}$$

where

$$\lambda_J = 1 - \frac{\lambda_{11}}{d}$$

Here λ_{11} is the smallest eigenvalue of A, given approximately by formula 4.7.3, and d is the (constant) diagonal of A,

$$d = \frac{2}{\Delta x^2} + \frac{2}{\Delta y^2} + a.$$

Table 4.7.1
SOR Iterations Required for Convergence

$a = -40$	(SOR diverged for all values of ω, Δx attempted)		
$a = 0$	ω	$\Delta x = 0.1$	$\Delta x = 0.05$	$\Delta x = 0.025$

$a = 0$	ω	$\Delta x = 0.1$	$\Delta x = 0.05$	$\Delta x = 0.025$
	1.00	124	492	1958
	1.05	112	446	1773
	1.10	101	404	1604
	1.15	91	365	1451
	1.20	82	329	1310
	1.25	73	296	1180
	1.30	65	266	1060
	1.35	57	238	948
	1.40	50	211	845
	1.45	42	186	748
	1.50	34	163	658
	1.55	27*	140	573
	1.60	30	119	493
	1.65	38	98	418
	1.70	41	75	346
	1.75	50	56*	276
	1.80	61	74	207
	1.85	84	89	127*
	1.90	125	134	160
	1.95	255	264	280

$a = 40$	ω	$\Delta x = 0.1$	$\Delta x = 0.05$	$\Delta x = 0.025$
	1.00	45	168	654
	1.05	40	152	593
	1.10	36	138	536
	1.15	33	124	485
	1.20	29	112	438
	1.25	25	101	395
	1.30	21	90	355
	1.35	20*	80	318
	1.40	21	71	283
	1.45	22	62	251
	1.50	24	53	220
	1.55	25	44	191
	1.60	28	41*	164
	1.65	35	43	138
	1.70	44	46	112
	1.75	48	51	83*
	1.80	66	69	84
	1.85	84	88	94
	1.90	123	130	161
	1.95	250	256	266

Table 4.7.2

	N	Experimental ω_{opt}	Number of iterations	
			$\omega = \omega_{opt}$	$\omega = 1$
$a = 0$	10	1.55	27	124
	20	1.75	56	492
	40	1.85	127	1958
$a = 40$	10	1.35	20	45
	20	1.60	41	168
	40	1.75	83	654

The formula 4.7.4 predicts values of ω_{opt} as shown below.

	N	ω_{opt}
$a = 0$	10	1.53
	20	1.73
	40	1.85
$a = 40$	10	1.33
	20	1.58
	40	1.76

These agree very well with the experimental ω_{opt} (Table 4.7.2).
When Δx and Δy are small and equal, 4.7.4 reduces to

$$\omega_{opt} \approx 2 - 2(\pi^2 + \tfrac{1}{2}a)^{1/2}\,\Delta x.$$

It has also been shown that when $\omega = \omega_{opt}$, for this problem, the successive over-relaxation error will decrease each iteration by about the factor

$$\lambda_{SOR} = \omega_{opt} - 1 \approx 1 - 2(\pi^2 + \tfrac{1}{2}a)^{1/2}\,\Delta x.$$

According to this theory, then, the number of iterations to reduce the error by a factor of 10^6 is I, where

$$(1 - 2(\pi^2 + \tfrac{1}{2}a)^{1/2}\,\Delta x)^I = 10^{-6}$$

or (for Δx small)

$$I \approx \frac{\ln(10^6)}{2(\pi^2 + \frac{1}{2}a)^{1/2} \Delta x} \approx \frac{6.91 \, N}{(\pi^2 + \frac{1}{2}a)^{1/2}}.$$

Thus for problem 4.7.1, at least, the $O(N)$ iteration count is predicted by the theory.

Unfortunately, we have a formula available for choosing ω optimally only for very simple problems. Some software, most notably the subroutines in ITPACK [Grimes et al., 1978], has been developed that chooses ω adaptively, as the calculations progress. Table 4.7.1 shows, however, that even if ω is not chosen "optimally," if it is chosen "reasonably" (between 1 and 2, and $2 - O(\Delta x)$ for large problems), "reasonably" good results can be expected.

Consider now the more difficult nonlinear partial differential equation

$$u_{xx} + u_{yy} + u_x + u_y - x^2 y^2 u^3 = 1,$$

with $u = 0$ on the boundary of the region shown below. By using the usual central differences, this is discretized by

$$\frac{U(x_{i+1},y_j) - 2U(x_i, y_j) + U(x_{i-1}, y_j)}{\Delta x^2}$$

$$+ \frac{U(x_i, y_{j+1}) - 2U(x_i, y_j) + U(x_i, y_{j-1})}{\Delta y^2}$$

$$+ \frac{U(x_{i+1}, y_j) - U(x_{i-1}, y_j)}{2 \, \Delta x} + \frac{U(x_i, y_{j+1}) - U(x_i, y_{j-1})}{2 \, \Delta y} \quad (4.7.5)$$

$$- x_i^2 y_j^2 U(x_i, y_j)^3 = 1.$$

This nonlinear system of equations, which has a nonsymmetric Jacobian matrix, could be solved using Newton's method, with SOR used to solve the linear system at each Newton iteration. However, it is much easier to simply use the nonlinear over-relaxation iteration 4.6.2 directly, and this is what was done.

Zero starting values were used, and since the exact solution is not known (it usually is not!), the NLOR iteration was stopped when the maximum change during an iteration was less than 10^{-8}. The number of iterations required for various values of ω and $\Delta x = \Delta y$ is shown in Table 4.7.3. The

Table 4.7.3
NLOR Iterations Required for Convergence

ω	$\Delta x = 0.2$	$\Delta x = 0.1$	$\Delta x = 0.05$
1.00	72	266	963
1.05	65	242	878
1.10	59	220	800
1.15	52	200	729
1.20	47	181	662
1.25	41	163	600
1.30	36	147	542
1.35	30	131	488
1.40	22*	116	437
1.45	25	102	389
1.50	28	88	344
1.55	32	74	300
1.60	36	59	259
1.65	41	45*	218
1.70	50	53	178
1.75	61	63	138
1.80	76	76	85*
1.85	106	103	113
1.90	157	161	153
1.95	320	315	304

NLOR iteration appears to work very well on this nonlinear, nonsymmetric problem. If ω is chosen optimally (marked with asterisks), the number of iterations again appears to be $O(N)$, doubling as N ($=2/\Delta x$) is doubled. Although the Jacobian is not positive-definite, or even symmetric, it is diagonal-dominant, and this is no doubt related to the success of NLOR on this problem (see Theorem 4.6.2).

The $O(\Delta x^2)$ accuracy of the central difference approximation is confirmed by calculating the error at a certain point (0.4, 0.4) for three values of

$\Delta x = \Delta y$, after convergence of the nonlinear equation solver (NLOR). The "exact" value $u(0.4, 0.4) = -0.13252$ was calculated accurately by using a finite element program, PDE/PROTRAN, which is discussed in Section 5.6. The experimental errors at this point are

$\Delta x = \Delta y$	$U(0.4, 0.4)$	Error
0.2	-0.12791	0.00461
0.1	-0.13159	0.00093
0.05	-0.13229	0.00023

The error is cut by a factor of 4 or 5 each time the stepsize is halved, as would be expected for a finite difference method with truncation error of second order in Δx and Δy.

The NLOR program used to solve 4.7.5, shown in Figure 4.7.2, is exceptionally simple despite the nonlinearity of the problem and the fact that the region is nonrectangular.

Finally, we consider the three-dimensional boundary value problem

$$u_{xx} + u_{yy} + u_{zz} = 1 \qquad \text{in the box } (0, 1) \times (0, 1) \times (0, 1)$$

with $$\frac{\partial u}{\partial n} = 0 \qquad \text{on the boundary } y = 0 \qquad (4.7.6)$$

and $$u = 0 \qquad \text{on the rest of the boundary.}$$

When the second derivatives are replaced by their usual second-order finite difference approximations, we get $(N - 1)^2 N$ linear equations, one centered on each of the unknowns $u(x_i, y_j, z_k)$, $1 \le i \le N - 1, 0 \le j \le N - 1, 1 \le k \le N - 1$. The first boundary condition is enforced by replacing $u(x_i, y_{-1}, z_k)$ with $u(x_i, y_1, z_k)$ where it appears in these linear equations. It is left as an exercise (Problem 9) to show that when the successive over-relaxation iterative method is used to solve this linear system, with ω chosen optimally, the number of iterations required is experimentally proportional to N for this three-dimensional problem.

Three level surfaces of the approximate solution of 4.7.6, calculated with $N = 12$, are shown in Figure 4.7.3. They are plotted using a special contour surface plotting program developed by the author [Sewell, 1988]. Because the contours are drawn using thick, opaque bands that mask the area in the

162   BOUNDARY VALUE PROBLEMS

```
      IMPLICIT DOUBLE PRECISION(A-H,O-Z)
C                             N = NUMBER OF POINTS IN X,Y DIRECTIONS
      PARAMETER (N=40)
      DIMENSION U(0:N,0:N),X(0:N),Y(0:N)
      DX = 2.D0/N
      DY = DX
C                             TRY VARIOUS VALUES OF THE PARAMETER OMEGA
      DO 30 IW=0,19
        W = 1.0 + IW*0.05
C                             SET UNKNOWNS AND BOUNDARY VALUES TO ZERO
        DO 5 I=0,N
        DO 5 J=0,N
          X(I) = -1.0 + I*DX
          Y(J) = -1.0 + J*DY
    5   U(I,J) = 0.0
C                             BEGIN NLOR ITERATION
        DO 15 ITER=1,10000
          DELMAX = 0.0
C                             UPDATE UNKNOWNS USING NLOR FORMULA
          DO 10 I=1,N-1
          LIM = 1
          IF (I.LE.N/2) LIM = N/2 + 1
          DO 10 J=LIM,N-1
            DELTA = W*( (U(I+1,J)-2*U(I,J)+U(I-1,J))/DX**2
     &                +(U(I,J+1)-2*U(I,J)+U(I,J-1))/DY**2
     &                    +(U(I+1,J)-U(I-1,J))/(2*DX)
     &                    +(U(I,J+1)-U(I,J-1))/(2*DY)
     &                - X(I)**2*Y(J)**2*U(I,J)**3 - 1.0 )
     &            / (-2/DX**2 - 2/DY**2 - X(I)**2*Y(J)**2*3*U(I,J)**2)
            U(I,J) = U(I,J) - DELTA
            DELMAX = MAX(DELMAX,ABS(DELTA))
   10     CONTINUE
          MAXITR = ITER
C                             STOP IF MAX CHANGE .LT. 10**(-8)
          IF (DELMAX.LT.1.D-8) GO TO 20
   15   CONTINUE
   20   PRINT 25, W,MAXITR
   25   FORMAT (5X,F8.2,I10)
   30 CONTINUE
      STOP
      END
```

Figure 4.7.2
FORTRAN77 Program Implementing NLOR Iteration

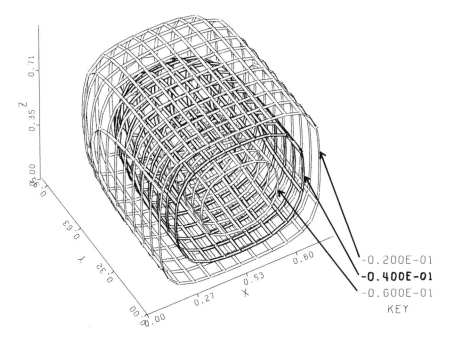

Figure 4.7.3
Three Level Surfaces of the Solution to $\nabla^2 u = 1$

immediate background, a sense of depth is preserved, yet the background contours are still visible between the bands. In the original plot, the three contour surfaces are drawn using contrasting colors.

As seen from the unsuccessful attempt to solve 4.7.1 when $a = -40$ (an indefinite problem), successive over-relaxation—as well as other iterative methods—is certainly not useful for as large a class of problems as direct linear equation solvers (based on Gaussian elimination). Nevertheless, SOR/NLOR has been observed to work well on many problems, such as 4.7.5, to which the known theory does not apply. And when it does work, it is usually faster than band solvers, at least when the number of unknowns is large.

While there are other, newer, iterative methods that are faster than successive over-relaxation for partial differential equation-type applications, and fast direct methods (see Section 0.5) are competitive with it in speed (and more robust), the extraordinary simplicity of successive over-relaxation ensures its continued popularity for many years to come, at least for "one-shot" programs.

4.8   The Conjugate-Gradient Method

Another popular iterative method for solving the linear systems that are generated by boundary value problems is the conjugate-gradient method. For the M by M linear system $A\mathbf{x} = \mathbf{b}$, where A is symmetric, the conjugate-gradient method is defined by the iteration

$$\mathbf{x_0} = \text{starting guess at solution}$$

$$\mathbf{r_0} = \mathbf{b} - A\mathbf{x_0}$$

$$\mathbf{p_0} = \mathbf{r_0}$$

$n = 1,\ldots$

$$
\begin{aligned}
&\lambda_n = (\mathbf{r_n}^T\mathbf{p_n})/(\mathbf{p_n}^T A\mathbf{p_n}) \\
&\mathbf{x_{n+1}} = \mathbf{x_n} + \lambda_n\mathbf{p_n} \\
&\mathbf{r_{n+1}} = \mathbf{r_n} - \lambda_n A\mathbf{p_n} \\
&\alpha_n = -(\mathbf{r_{n+1}}^T A\mathbf{p_n})/(\mathbf{p_n}^T A\mathbf{p_n}) \\
&\mathbf{p_{n+1}} = \mathbf{r_{n+1}} + \alpha_n\mathbf{p_n}.
\end{aligned}
$$
                                                                    (4.8.1)

First, note that $\mathbf{r_n}$ is the residual $\mathbf{b} - A\mathbf{x_n}$, as can be proved by induction. For $n = 0$, this is true by the definition of $\mathbf{r_0}$. Assume that it is true for n, that is, that $\mathbf{r_n} = \mathbf{b} - A\mathbf{x_n}$, and we verify that

$$\mathbf{r_{n+1}} = \mathbf{r_n} - \lambda_n A\mathbf{p_n} = \mathbf{b} - A\mathbf{x_n} - A\lambda_n\mathbf{p_n} = \mathbf{b} - A\mathbf{x_{n+1}}.$$

If A is also positive-definite, we can show that this residual (which is a measure of the error in $\mathbf{x_n}$) decreases with every iteration, at least in the norm defined by

$$\|\mathbf{e}\| \equiv (\mathbf{e}^T A^{-1}\mathbf{e})^{1/2}.$$

Since A, and therefore also A^{-1}, is positive-definite, this is a legitimate norm and equal to zero only when $\mathbf{e} = \mathbf{0}$.

Then in this norm (recall that $A^T = A$),

$$
\begin{aligned}
\|\mathbf{r_{n+1}}\|^2 &= (\mathbf{r_n} - \lambda_n A\mathbf{p_n})^T A^{-1}(\mathbf{r_n} - \lambda_n A\mathbf{p_n}) \\
&= \mathbf{r_n}^T A^{-1}\mathbf{r_n} - 2\lambda_n\mathbf{r_n}^T\mathbf{p_n} + \lambda_n^2\mathbf{p_n}^T A\mathbf{p_n} \\
&= \|\mathbf{r_n}\|^2 - 2\lambda_n\mathbf{r_n}^T\mathbf{p_n} + \lambda_n\mathbf{r_n}^T\mathbf{p_n} \\
&= \|\mathbf{r_n}\|^2 - (\mathbf{r_n}^T\mathbf{p_n})^2/(\mathbf{p_n}^T A\mathbf{p_n}).
\end{aligned}
$$
                                                                    (4.8.2)

This shows that the residual is nonincreasing, at least. To show that the residual actually decreases (unless it is already zero) each step, we need the following results, which can be proved by induction [Sewell, 1985, Appendix 1].

$$\mathbf{r}_i^T \mathbf{p}_j = 0 \qquad \text{for } i > j, \tag{4.8.3a}$$

$$\mathbf{p}_i^T A \mathbf{p}_j = 0 \qquad \text{for } i \neq j. \tag{4.8.3b}$$

Taking the scalar product of both sides of $\mathbf{p_n} = \mathbf{r_n} + \alpha_{n-1}\mathbf{p_{n-1}}$ with $\mathbf{r_n}$, and using 4.8.3a gives

$$\mathbf{r_n}^T \mathbf{p_n} = \mathbf{r_n}^T \mathbf{r_n}. \tag{4.8.4}$$

Now 4.8.2 becomes, using 4.8.4

$$\|\mathbf{r_{n+1}}\|^2 = \|\mathbf{r_n}\|^2 - \frac{(\mathbf{r_n}^T \mathbf{r_n})^2}{(\mathbf{p_n}^T A \mathbf{p_n})}, \tag{4.8.5}$$

which proves that the residual decreases at each step (in the A^{-1} norm) unless it is already zero, in which case we have the exact solution, $\mathbf{x_n}$.

Notice that when A is positive-definite, the term $\mathbf{p_n}^T A \mathbf{p_n}$ that appears in the denominators in 4.8.1 and 4.8.5 is nonnegative and can be zero only if $\mathbf{p_n} = \mathbf{0}$, but 4.8.4 shows that this can only happen if we already have the exact solution, that is, if $\mathbf{r_n} = \mathbf{0}$. Thus, when A is positive-definite, the conjugate-gradient method cannot "break-down" before convergence.

The proof of 4.8.3 does not require A to be positive-definite; it is only required that A be symmetric and that premature breakdown not occur. Thus, barring the unlikely event of an exact breakdown before convergence, we can see that the vectors $\mathbf{p_0}, \ldots, \mathbf{p_{M-1}}$ (M is the rank of A) are linearly independent. For, if $\Sigma c_i \mathbf{p_i} = \mathbf{0}$, then taking the scalar product of both sides with $A\mathbf{p_j}$ gives $c_j = 0$, using 4.8.3b and the assumption that $\mathbf{p_j}^T A \mathbf{p_j} \neq 0$. Then 4.8.3a shows that $\mathbf{r_M}$ is orthogonal to these M independent vectors and is therefore exactly zero. So, unless it breaks down prematurely, the conjugate-gradient method is guaranteed to give the exact solution in at most M iterations, *assuming the calculations are done exactly*.

Thus the conjugate-gradient method would seem to combine the best features of direct and iterative methods. If A is positive-definite, the solution improves with each iteration and is guaranteed to be exact within M iterations, and even if A is symmetric but indefinite, the solution is guaranteed to be exact within M iterations, barring an accident of probability zero. Unfortunately, the guaranteed convergence property is not of much practical significance even in the positive-definite case. If A is well conditioned, convergence occurs much sooner than M iterations for the type of problems we are interested in, while if A is ill conditioned, x_M may be very far from exact due to roundoff error.

We shall now use the conjugate-gradient method 4.8.1 to solve the linear system 4.7.1, previously solved by SOR. It was shown earlier that this system is positive-definite provided a is greater than about -19.74, and so the conjugate-gradient method is theoretically guaranteed to work for such values of a.

The FORTRAN77 program used to solve 4.7.1 with the conjugate-gradient method is shown in Figure 4.8.1. Note that, as when the successive over-relaxation iterative method was used (Figure 4.7.1), the solution $\mathbf{u}$ ($=\mathbf{x}$) and all other vectors are stored in two-dimensional arrays for convenience. The matrix A is never explicitly formed or stored; it appears only in the matrix-vector product $A\mathbf{p}$. This product is most conveniently calculated as $\mathbf{b}$ minus the residual when $\mathbf{p}$ is substituted into the linear equations, that is, $A\mathbf{p} = \mathbf{b} - (\mathbf{b} - A\mathbf{p})$.

```
      IMPLICIT DOUBLE PRECISION(A-H,O-Z)
C                              N = NUMBER OF POINTS IN X,Y DIRECTIONS
C                              A = PDE PARAMETER
      PARAMETER (N=40, A=40.0)
      DIMENSION U(0:N,0:N),R(0:N,0:N),P(0:N,0:N),
     &          AP(0:N,0:N),B(0:N,0:N),X(0:N),Y(0:N)
      DOUBLE PRECISION LAMBDA
      DX = 1.D0/N
      DY = DX
C                              U = 0
      DO 5 I=1,N-1
      DO 5 J=1,N-1
    5 U(I,J) = 0.0
C                              SET VALUES OF U,P ON BOUNDARY (THESE WILL
C                              NEVER CHANGE)
      DO 10 L=0,N
         X(L) = L*DX
         Y(L) = L*DY
         U(L,0) = 0.0
         P(L,0) = 0.0
         U(L,N) = X(L)
         P(L,N) = X(L)
         U(0,L) = 0.0
         P(0,L) = 0.0
         U(N,L) = Y(L)**3
         P(N,L) = Y(L)**3
   10 CONTINUE
C                              R = B - A*U
C                              B = R   (SINCE U=0)
```

```
C                                  P = R
      DO 15 I=1,N-1
      DO 15 J=1,N-1
        R(I,J) = (-U(I+1,J)+2*U(I,J)-U(I-1,J))/DX**2
     &          +(-U(I,J+1)+2*U(I,J)-U(I,J-1))/DY**2
     &          + A*U(I,J) - X(I)*Y(J)*(A*Y(J)**2-6)
        B(I,J) = R(I,J)
        P(I,J) = R(I,J)
   15 CONTINUE
C                                  BEGIN CONJUGATE GRADIENT ITERATION
      DO 30 ITER=1,1000
C                                  AP = B - (B-A*P)
      DO 20 I=1,N-1
      DO 20 J=1,N-1
        AP(I,J) = B(I,J) - ( (-P(I+1,J)+2*P(I,J)-P(I-1,J))/DX**2
     &                      +(-P(I,J+1)+2*P(I,J)-P(I,J-1))/DY**2
     &                      + A*P(I,J) - X(I)*Y(J)*(A*Y(J)**2-6) )
   20   CONTINUE
C                                  LAMBDA = R*P / P*AP
      LAMBDA = DOT(R,P,N)/DOT(P,AP,N)
C                                  U = U + LAMBDA*P
      CALL UPDATE(U,U,LAMBDA,P,N)
C                                  R = R - LAMBDA*AP
      CALL UPDATE(R,R,-LAMBDA,AP,N)
C                                  ALPHA = -R*AP / P*AP
      ALPHA = -DOT(R,AP,N)/DOT(P,AP,N)
C                                  P = R + ALPHA*P
      CALL UPDATE(P,R,ALPHA,P,N)
C                                  CALCULATE MAXIMUM ERROR
      ERMAX = 0.0
      DO 25 I=1,N-1
      DO 25 J=1,N-1
        ERR = ABS(U(I,J) - X(I)*Y(J)**3)
        ERMAX = MAX(ERMAX,ERR)
   25   CONTINUE
      MAXITR = ITER
C                                  STOP IF ERROR .LT. 10**(-6)
      IF (ERMAX.LT.1.D-6) GO TO 35
   30 CONTINUE
   35 PRINT 40,MAXITR
   40 FORMAT (I10)
      STOP
      END
      FUNCTION DOT(X,Y,N)
      IMPLICIT DOUBLE PRECISION(A-H,O-Z)
      DIMENSION X(0:N,0:N),Y(0:N,0:N)
C                                  CALCULATE DOT PRODUCT OF X AND Y
      DOT = 0.0
      DO 5 I=1,N-1
      DO 5 J=1,N-1
```

(continued)

168   BOUNDARY VALUE PROBLEMS

```
   5 DOT = DOT + X(I,J)*Y(I,J)
     RETURN
     END
     SUBROUTINE UPDATE(X,Y,S,Z,N)
     IMPLICIT DOUBLE PRECISION(A-H,O-Z)
     DIMENSION X(Ø:N,Ø:N),Y(Ø:N,Ø:N),Z(Ø:N,Ø:N)
C                              CALCULATE X = Y + S*Z
     DO 5 I=1,N-1
     DO 5 J=1,N-1
   5 X(I,J) = Y(I,J) + S*Z(I,J)
     RETURN
     END
```

Figure 4.8.1
FORTRAN77 Program Implementing Conjugate Gradient Iteration

Table 4.8.1 shows the number of iterations required to reduce the maximum error to less than 10^{-6} (with zero starting values used) for each set of parameters $(a, \Delta x = \Delta y)$. If the exact solution were not known, the residual could be used to determine when to stop the iteration.

The number of iterations appears to be proportional to N $(= 1/\Delta x)$, even when $a = -40$ and the matrix is not positive-definite. The amount of work per iteration is $O(N^2)$, because the matrix-vector multiplication $A\mathbf{p_n}$ is the most time-consuming calculation in each iteration, and, since there are at most five nonzeros per row in A, this requires work proportional to the order of the matrix, $M = (N - 1)^2$ (assuming advantage is taken of the sparsity of A in this multiplication). This means that, experimentally, the total work to solve 4.7.1 by using the conjugate-gradient method is $O(N^3)$, the same as for successive over-relaxation with optimal parameter choice, and much better than the $O(N^4)$ for the band solver. But now we no longer have to worry about choosing any parameters optimally.

For both the successive over-relaxation and the conjugate-gradient methods (and most other iterative methods), the total storage is $O(N^2)$ for this two-dimensional problem, since (at most) only the nonzeros of A have to be stored. This compares with $O(N^3)$ for a band solver.

Table 4.8.1
Conjugate-Gradient Iterations Required for Convergence

	$\Delta x = 0.1$	$\Delta x = 0.05$	$\Delta x = 0.025$
$a = -40$	30	63	128
$a = 0$	24	50	101
$a = 40$	20	41	84
M (# unknowns)	81	361	1521

For three-dimensional problems like 4.7.6, the successive over-relaxation (see Problem 9) and the conjugate-gradient methods both require $O(N^4)$ work for an $N \times N \times N$ finite difference grid, which compares *very* favorably with the $O(N^7)$ work required by the band solver. When the problem is one-dimensional, on the other hand, no iterative method can beat the band (usually tridiagonal) solver.

It has been shown [Hageman and Young, 1981, Section 7.3] that, for positive-definite systems, the number of iterations required by the conjugate-gradient method is proportional to the square root of the "spectral condition number," that is, the ratio of the largest eigenvalue of A to the smallest. The eigenvalues of the matrix of 4.7.1 have already been found; they are given in 4.7.2. Thus the spectral condition number is ($\Delta x = \Delta y = 1/N$),

$$\frac{\lambda_{max}}{\lambda_{min}} = \frac{\lambda_{N-1,N-1}}{\lambda_{11}} \approx \frac{8N^2 + a}{2\pi^2 + a} = O(N^2). \tag{4.8.6}$$

The number of iterations, according to this theory, should thus be proportional to N, as observed experimentally. From 4.8.6 it can also be seen that increasing a should decrease the condition number somewhat, resulting in faster convergence. This is observable experimentally in Table 4.8.1.

The fact that decreasing the condition number speeds convergence also explains why there is so much interest in "preconditioning" the matrix, that is, multiplying $A\mathbf{x} = \mathbf{b}$ by a positive-definite matrix B^{-1} such that $B^{-1}A$ has a lower condition number than A, before applying the conjugate gradient method. Although $B^{-1}A$ may no longer be symmetric, it is possible to take care of this problem with a minor modification to 4.8.1 (see Problem 11).

Although we have assumed A to be symmetric, there is nothing to prevent us from applying the conjugate-gradient iteration (4.8.1) to a nonsymmetric linear system. Therefore, we made 4.7.1 nonsymmetric by adding $u_x - y^3$ to the partial differential equation, using central differences to approximate u_x. With $a = 40$, the nonsymmetric term hardly slowed the convergence at all, but with $a = -40$, the iteration diverged. Thus, like the successive over-relaxation iteration, the conjugate-gradient method actually solves a much larger class of problems than the known theory would indicate.

There also exists a generalization of the conjugate-gradient algorithm that works well on a much wider range of nonsymmetric problems, called the Lanczos, or biconjugate gradient method [Sewell, 1985, Section 3.4]. In any case, however, all second-order linear boundary value problems in one variable and many multi-dimensional boundary value problems can be put into a form that can be approximated by using symmetric finite difference equations, as shown in Problem 10.

4.9   Systems of Differential Equations

All of the steady-state problems studied thus far have involved only a single partial differential equation. However, generalization of the techniques discussed to a system of m partial differential equations is very straightforward. Consider, for example, the following system of two ordinary differential equations:

$$au_{xx} + bv_{xx} = f(x),$$

$$cu_{xx} + dv_{xx} = g(x)$$

with

$$u(0) = u(1) = 0,$$

$$v(0) = v(1) = 0.$$

If the second derivatives are approximated by using the usual second-order formula (1.2.5b), with $x_i = i\,\Delta x$, $\Delta x = 1/N$, we get $2(N-1)$ finite difference equations for the $2(N-1)$ unknowns. In order to minimize the bandwidth (in case the linear system is to be solved by a band solver), the unknowns and equations are ordered in an alternating fashion. The system then looks like

$$
\begin{bmatrix}
-2a & -2b & a & b & 0 & 0 & \cdot & \cdot \\
-2c & -2d & c & d & 0 & 0 & \cdot & \cdot \\
 & & & & & & & \\
a & b & -2a & -2b & a & b & \cdot & \cdot \\
c & d & -2c & -2d & c & d & \cdot & \cdot \\
 & & & & & & & \\
0 & 0 & a & b & -2a & -2b & \cdot & \cdot \\
0 & 0 & c & d & -2c & -2d & \cdot & \cdot \\
 & & & & & & & \\
\cdot & \cdot & \cdot & \cdot & \cdot & \cdot & \cdot & \cdot \\
\cdot & \cdot & \cdot & \cdot & \cdot & \cdot & \cdot & \cdot \\
 & & & & & & & \\
 & & & & & \cdot & \cdot & -2a & -2b \\
 & & & & & \cdot & \cdot & -2c & -2d
\end{bmatrix}
\begin{bmatrix}
U(x_1) \\
V(x_1) \\
\\
U(x_2) \\
V(x_2) \\
\\
U(x_3) \\
V(x_3) \\
\\
\cdot \\
\\
U(x_{N-1}) \\
V(x_{N-1})
\end{bmatrix}
=
\begin{bmatrix}
\Delta x^2 f(x_1) \\
\Delta x^2 g(x_1) \\
\\
\Delta x^2 f(x_2) \\
\Delta x^2 g(x_2) \\
\\
\Delta x^2 f(x_3) \\
\Delta x^2 g(x_3) \\
\\
\cdot \\
\\
\Delta x^2 f(x_{N-1}) \\
\Delta x^2 g(x_{N-1})
\end{bmatrix}
.
$$

This system is no longer tridiagonal, but it still has a small bandwidth, and the storage and CPU time required to solve it are still $O(N)$. The linear system is "block tridiagonal," that is, it can be thought of as a tridiagonal linear system whose "elements" are 2×2 matrices. In two and three dimensions, a typical second-order system of m partial differential equations may generate linear systems as shown in Figures 4.5.2 and 4.5.3, respectively, except that now each "element" consists of an $m \times m$ matrix block.

As an example, consider a rigid elastic beam clamped at one end and simply supported at the other, as drawn below.

The vertical displacement $u(x)$ of the beam satisfies the *fourth-order* boundary value problem

$$u_{xxxx} = 0,$$

$$u(0) = 0, \qquad u(1) = 1, \qquad\qquad (4.9.1)$$

$$u_x(0) = 0, \qquad u_{xx}(1) = 0.$$

This can be reformulated as a system of two second-order differential equations by introducing the auxiliary variable $v \equiv u_{xx}$:

$$u_{xx} - v = 0, \qquad u(0) = 0, \qquad u(1) = 1,$$

$$v_{xx} = 0, \qquad u_x(0) = 0, \qquad v(1) = 0.$$

This is approximated by the finite-difference equations

$$\frac{U(x_{i+1}) - 2U(x_i) + U(x_{i-1})}{\Delta x^2} - V(x_i) = 0,$$

$$\frac{V(x_{i+1}) - 2V(x_i) + V(x_{i-1})}{\Delta x^2} = 0, \qquad\qquad \text{for } i = 1, \ldots, N - 1. \quad (4.9.2)$$

A problem with 4.9.2 is that it assumes that $U(x_0)$ and $V(x_0)$ are both known, when in fact only $U(x_0)$ is known. However, the boundary conditions on u at $x = 0$ can be approximated by $U(x_0) = 0$, $U(x_{-1}) = U(x_1)$, and these can be used in conjunction with the equation

$$\frac{U(x_1) - 2U(x_0) + U(x_{-1})}{\Delta x^2} - V(x_0) = 0$$

to find $V(x_0)$:

$$V(x_0) = \frac{2U(x_1)}{\Delta x^2}.$$

Since also

$$U(x_0) = 0,$$
$$V(x_N) = 0,$$
$$U(x_N) = 1,$$

the $2(N-1)$ finite difference equations 4.9.2 can now be used to calculate the $2(N-1)$ unknowns $U(x_1), V(x_1), \ldots, U(x_{N-1}), V(x_{N-1})$. The linear system has the form

$$
\begin{bmatrix}
-2 & -\Delta x^2 & 1 & 0 & 0 & 0 & & \cdot & \cdot \\
2/\Delta x^2 & -2 & 0 & 1 & 0 & 0 & & \cdot & \cdot \\
& & & & & & & & \\
1 & 0 & -2 & -\Delta x^2 & 1 & 0 & & \cdot & \cdot \\
0 & 1 & 0 & -2 & 0 & 1 & & \cdot & \cdot \\
& & & & & & & & \\
0 & 0 & 1 & 0 & -2 & -\Delta x^2 & & \cdot & \cdot \\
0 & 0 & 0 & 1 & 0 & -2 & & \cdot & \cdot \\
& & & & & & & & \\
\cdot & \cdot & \cdot & \cdot & & \cdot & & \cdot & \\
\cdot & \cdot & \cdot & \cdot & & \cdot & & \cdot & \\
& & & & & & & & \\
\cdot & \cdot & \cdot & \cdot & \cdot & \cdot & -2 & -\Delta x^2 \\
\cdot & \cdot & & \cdot & \cdot & \cdot & 0 & -2 \\
\end{bmatrix}
\begin{bmatrix}
U(x_1) \\
V(x_1) \\
\\
U(x_2) \\
V(x_2) \\
\\
U(x_3) \\
V(x_3) \\
\\
\cdot \\
\cdot \\
\\
U(x_{N-1}) \\
V(x_{N-1})
\end{bmatrix}
=
\begin{bmatrix}
0 \\
0 \\
\\
0 \\
0 \\
\\
0 \\
0 \\
\\
\cdot \\
\cdot \\
\\
-1 \\
0
\end{bmatrix}
$$

Although this banded linear system, with half-bandwidth $L = 2$, is neither diagonal-dominant nor positive-definite, the band solver of Figure 0.4.3, which does no pivoting, was used successfully to solve this system. The FORTRAN77 program is shown in Figure 4.9.1 below.

The maximum errors in U (exact solution is $u(x) = -0.5x^3 + 1.5x^2$) were $0.38481E - 4$ and $0.96217E - 5$ when Δx was set to 0.02 and 0.01, respectively. Since $0.38481E - 4/0.96217E - 5 = 3.999$, the expected $O(\Delta x^2)$ convergence rate is clearly observed. Since the fourth derivatives of u and v are identically zero, central differences yield exact approximations to their

```
      IMPLICIT DOUBLE PRECISION(A-H,O-Z)
C                         N = NUMBER OF GRID POINTS
      PARAMETER (N=100)                  .
      DIMENSION A(2*N-2,5),U(2*N-2),B(2*N-2)
      DX = 1.D0/N
C                         THE COLUMNS OF A ARE THE DIAGONALS OF THE
C                         COEFFICIENT MATRIX. B IS THE RIGHT HAND SIDE.
      M = 2*(N-1)
      DO 5 I=1,M
        B(I) = 0.0
        A(I,1) =  1.0
        A(I,2) =  0.0
        A(I,3) = -2.0
        A(I,4) =  0.0
        IF (MOD(I,2).EQ.1) A(I,4) = -DX**2
        A(I,5) =  1.0
    5 CONTINUE
      B(M-1) = -1.0
      A(2,2) = 2/DX**2
C                         CALL LBAND
      L = 2
      CALL LBAND(A,U,B,M,M,L)
C                         CALCULATE MAXIMUM ERROR IN U (ODD COMPONENTS
C                         OF SOLUTION VECTOR)
      ERMAX = 0.0
      DO 10 I=1,N-1
        XI = I*DX
        SOL = -0.5*XI**3 + 1.5*XI**2
        ERMAX = MAX(ERMAX,ABS(U(2*I-1)-SOL))
   10 CONTINUE
      PRINT 15,ERMAX
   15 FORMAT (E15.5)
      STOP
      END
```

Figure 4.9.1
FORTRAN77 Program to Solve Elastic Beam Problem

second derivatives (see 1.2.5b). The fact that 4.9.2 does not produce exact answers is due to the inexactness of the treatment of the boundary condition $u_x(0) = 0$ (see 1.2.5a).

4.10   The Eigenvalue Problem

Consider the one-dimensional eigenvalue problem (cf. 4.0.5)

$$u_{xx} - v(x)u_x + a(x)u = \lambda\rho(x)u,$$
$$u(0) = 0, \qquad\qquad (4.10.1)$$
$$u(1) = 0,$$

which has been normalized by dividing by the coefficient of u_{xx}. Replacing u_x and u_{xx} by their usual centered difference approximations (1.2.5a, b) leads to a generalized matrix eigenvalue problem of the form $A\mathbf{u} = \lambda B\mathbf{u}$, where $\mathbf{u} = (U(x_1), \ldots, U(x_{N-1}))$ and B is a diagonal matrix. However, the matrix A is not symmetric, unless $v \equiv 0$, and for both theoretical and practical reasons it is preferable that A be symmetric. Fortunately, the first derivative term in 4.10.1 can be eliminated by multiplying by $e^{-V(x)}$, where $V(x)$ is any indefinite integral of $v(x)$, that is, $V_x(x) = v(x)$:

$$(u_x e^{-V(x)})_x + a(x)e^{-V(x)}u = \lambda\rho(x)e^{-V(x)}u.$$

Therefore, we will assume that our eigenvalue problem has been put into the "Sturm-Liouville" form

$$(D(x)u_x)_x + a(x)u = \lambda\rho(x)u,$$
$$u(0) = 0, \qquad\qquad (4.10.2)$$
$$u(1) = 0,$$

where $\rho(x)$ is strictly positive (if $\rho(x) < 0$, multiply by -1).

This eigenvalue problem can be approximated by the following *symmetric* centered difference formula ($x_i = i\Delta x$, $\Delta x = 1/N$):

$$\frac{D(x_{i+1/2})\left[\dfrac{U(x_{i+1}) - U(x_i)}{\Delta x}\right] - D(x_{i-1/2})\left[\dfrac{U(x_i) - U(x_{i-1})}{\Delta x}\right]}{\Delta x}$$
$$+ a(x_i)U(x_i) = \lambda\rho(x_i)U(x_i), \qquad 1 \le i \le N - 1,$$
$$U(x_0) = 0, \qquad\qquad (4.10.3)$$
$$U(x_N) = 0.$$

The truncation error for 4.10.3 is calculated to be

$$T_i = \tfrac{1}{24} h^2 \{[D(x_i)u_{xxx}(x_i)]_x + [D(x_i)u_x(x_i)]_{xxx}\} + O(h^3).$$

The differential equation eigenvalue problem 4.10.2 has now been approximated by a generalized matrix eigenvalue problem of the form

$$A\mathbf{u} = \lambda B\mathbf{u}, \qquad\qquad (4.10.4)$$

where A is tridiagonal with elements

$$A_{i,i+1} = \left(\frac{1}{\Delta x^2}\right) D(x_{i+1/2}),$$

$$A_{i,i} = -\left(\frac{1}{\Delta x^2}\right) [D(x_{i+1/2}) + D(x_{i-1/2})] + a(x_i),$$

$$A_{i,i-1} = \left(\frac{1}{\Delta x^2}\right) D(x_{i-1/2}),$$

and B is diagonal with elements $B_{i,i} = \rho(x_i)$. That A is symmetric is seen by observing that $A_{i+1,i} = (1/\Delta x^2)D(x_{i+1/2}) = A_{i,i+1}$.

The generalized matrix eigenvalue problem $A\mathbf{u} = \lambda B\mathbf{u}$ is equivalent to the standard eigenvalue problem $B^{-1}A\mathbf{u} = \lambda\mathbf{u}$. Unfortunately, $B^{-1}A$ is not, in general, symmetric. Techniques for finding the eigenvalues of symmetric matrices are more efficient and less complicated than those designed for nonsymmetric matrices, which must assume that the eigenvalues and eigenvectors are complex.

However, a standard *symmetric* eigenvalue problem with the same eigenvalues as $A\mathbf{u} = \lambda B\mathbf{u}$ can be constructed as follows: Since B is positive-definite (it is diagonal with positive elements), according to Theorem 0.3.1 it has a Cholesky decomposition $B = LL^T$, where L is a nonsingular lower triangular matrix. Of course, in this case, $L = L^T = B^{1/2}$, but in Section 5.10 an eigenproblem of the form $A\mathbf{u} = \lambda B\mathbf{u}$, where B is positive-definite but not diagonal, will be encountered, and therefore we want to consider the more general case for future reference. Thus,

$$A\mathbf{u} = \lambda LL^T\mathbf{u}$$

or

$$H\mathbf{v} = \lambda\mathbf{v}, \qquad\qquad (4.10.5)$$

where $H = L^{-1}AL^{-T}$ and $\mathbf{v} = L^T\mathbf{u}$. H is symmetric, since $H^T = (L^{-1}AL^{-T})^T = L^{-1}AL^{-T} = H$, so the eigenvalues λ are all real. Since B (and thus L^{-1}) is diagonal, H is, like A, tridiagonal. If B is not diagonal (as in Section 5.10), H may be a full matrix.

There are many readily available subroutines, such as the EISPACK routine TQL2 [Smith et al., 1974], which can be used to calculate efficiently the eigenvalues and eigenvectors of the symmetric tridiagonal matrix H. The eigenvectors of $A\mathbf{u} = \lambda B\mathbf{u}$ are then $\mathbf{u}_i = L^{-T}\mathbf{v}_i$, where the $\mathbf{v}_i$ are the eigenvectors of H.

Perhaps the most straightforward way to calculate some eigenvalues of a tridiagonal matrix is simply to use a good nonlinear equation solver to find some roots of $f(\lambda) \equiv \det(H - \lambda I) = 0$ (when this determinant is zero, $H - \lambda I$ is singular, so that $H\mathbf{v} - \lambda\mathbf{v} = \mathbf{0}$ has a nonzero solution) by evaluating $f(\lambda)$ directly from

$$
\det(H - \lambda I) = \begin{bmatrix} b_1 - \lambda & c_1 \\ a_2 & b_2 - \lambda & c_2 \\ & \vdots & \vdots & \vdots \\ & & a_k & b_k - \lambda & c_k \\ & & & a_{k+1} & b_{k+1} - \lambda & c_{k+1} \\ & & & & \vdots & \vdots \end{bmatrix}
$$

This determinant can be evaluated by defining $f_k(\lambda)$ to be the determinant of the $k \times k$ submatrix in the upper left-hand corner of $H - \lambda I$, and by using the recurrence relation

$$
f_1(\lambda) = b_1 - \lambda,
$$
$$
f_2(\lambda) = (b_2 - \lambda)f_1(\lambda) - a_2 c_1,
$$
$$
\vdots \qquad \vdots
$$
$$
f_{k+1}(\lambda) = (b_{k+1} - \lambda)f_k(\lambda) - a_{k+1}c_k f_{k-1}(\lambda),
$$
$$
\vdots \qquad \vdots
$$

to calculate $f(\lambda) = f_{N-1}(\lambda)$ for any given value of λ. In this case, since H is symmetric, $a_{k+1} = c_k$, but this same technique can still be used in the nonsymmetric case.

The work required to calculate $\det(H - \lambda I)$ is thus only $O(N)$. The root finder can therefore afford to do many more function evaluations than it could if H were a full matrix, since $O(N^3)$ work per evaluation is required in that case.

For multi-dimensional problems, the matrices A and H are banded but not tridiagonal. In this case, a good eigenvalue subroutine (e.g., EISPACK routine BANDR) will first reduce H to a symmetric tridiagonal matrix T through an orthogonal similarity transformation of the form $SHS^{-1} = T$, where $S^{-1} = S^T$, which preserves the eigenvalues of H (if $H\mathbf{z} = \lambda \mathbf{z}$, then $T(S\mathbf{z}) = \lambda(S\mathbf{z})$).

As an example, consider the membrane eigenvalue problem

$$-u_{xx} - u_{yy} = \lambda u \qquad \text{in } (0, 1) \times (0, 1),$$
$$u = 0 \qquad \text{on the boundary.}$$

$(4.10.6)$

This is discretized in the usual manner ($x_i = i\,\Delta x$, $y_j = j\,\Delta y$):

$$\frac{-U(x_{i+1}, y_j) + 2U(x_i, y_j) - U(x_{i-1}, y_j)}{\Delta x^2}$$

$$+ \frac{-U(x_i, y_{j+1}) + 2U(x_i, y_j) - U(x_i, y_{j-1})}{\Delta y^2} = \lambda U(x_i, y_j). \quad (4.10.7)$$

This discretization yields a matrix eigenvalue problem of the form $A\mathbf{u} = \lambda B\mathbf{u}$ with A symmetric (since there are no first-order terms) and $B = I$. With $\Delta x = \Delta y = 0.2$, the matrix A (times Δx^2) is identical to the symmetric, banded matrix exhibited in Figure 4.5.2.

EISPACK routines BANDR and TQL2 were used to calculate the eigenvalues of the symmetric band matrix A, with $\Delta x = \Delta y = 0.2$ and $\Delta x = \Delta y = 0.1$. The eigenvalues of 4.10.7 thus calculated are compared with the exact eigenvalues of 4.10.6, namely $\lambda_{kl} = (k^2 + l^2)\pi^2$, $k,l = 1, 2, \ldots$, in Table 4.10.1. (Note that the exact values of the *discrete* eigenvalues have already been calculated, in formula 4.7.2 with $a = 0$.)

The approximations to the smallest eigenvalues appear to have $O(\Delta x^2)$ accuracy. For example, the error in the first eigenvalue decreases by a factor of 4.3 when $\Delta x = \Delta y$ is decreased by a factor of 2.

It is clear from Table 4.10.1 that only the lowest eigenvalues are reasonable approximations to eigenvalues of the partial differential equation eigenvalue problem. Since the partial differential equation eigenproblem has an infinite number of eigenvalues, and A has only a finite number, it is not surprising that the higher eigenvalues are not accurately estimated! Since it is also true that in applications usually only the smallest, or smallest few, eigenvalues are of interest anyway, perhaps a technique that only finds one eigenvalue at a time, such as the inverse power method, is appropriate.

Table 4.10.1
Approximate and Exact Eigenvalues of Membrane
Problem

$\lambda(\Delta x = \Delta y = 0.2)$	$\lambda(\Delta x = \Delta y = 0.1)$	$\lambda(\text{exact})$
19.10	19.59	19.74
44.10	48.00	49.35
44.10	48.00	49.35
69.10	76.41	78.96
75.00	92.24	98.70
75.00	92.24	98.70
100.00	120.64	128.31
100.00	120.64	128.31
100.00	147.98	167.78
100.00	147.98	167.78
125.00	164.88	177.65
125.00	176.39	197.39
130.90	176.39	197.39
155.90	209.78	246.74
155.90	209.78	246.74
180.90	220.63	256.61
	220.63	256.61
	$\vdots$	$\vdots$
	780.42	
		∞

4.11 The Inverse Power Method

The inverse power method, which finds the smallest eigenvalue in absolute value, λ_1, of the matrix eigenvalue problem $A\mathbf{u} = \lambda B\mathbf{u}$ (cf. 4.10.4), is defined by the following iteration:

$$A\mathbf{u}_{n+1} = B\mathbf{u}_n$$

$$\mu_n = (\mathbf{u}_n^T B\mathbf{u}_n)/(\mathbf{u}_n^T B\mathbf{u}_{n+1}).$$
(4.11.1)

If λ_1 is very large or very small, it may be necessary to normalize $\mathbf{u}_n$ (replace $\mathbf{u}_n$ by $\mathbf{u}_n/\|\mathbf{u}_n\|_\infty$) before each iteration to prevent underflow or overflow. Notice also that since each step of the iteration 4.11.1 involves the solution of a linear system with the same coefficient matrix (A), it is clearly advantageous to form the LU decomposition of A (see Section 0.3) at the first step and to use this to solve the system on subsequent steps.

The following theorem gives conditions sufficient to ensure convergence of the inverse power method.

Theorem 4.11.1 Assume A and B are symmetric, B is positive-definite, A is nonsingular, and the eigenproblem $A\mathbf{u} = \lambda B\mathbf{u}$ has an eigenvalue λ_1 that is smaller in absolute value than all other eigenvalues. Then μ_n converges to λ_1 and $\lambda_1^n \mathbf{u}_n$ converges to a corresponding eigenvector (hence $\mathbf{u}_n$ is an approximate eigenvector, for large n), for "almost any" choice of $\mathbf{u}_0$.

Proof Recall that the eigenvalues of the symmetric matrix $H = L^{-1}AL^{-T}$ in 4.10.5 are the same as those of the eigenvalue problem $A\mathbf{u} = \lambda B\mathbf{u}$, where $B = LL^T$. By a fundamental theorem of linear algebra, there exists a unitary matrix S ($S^T = S^{-1}$) such that $SHS^{-1} = E$, where E is a diagonal matrix whose diagonal entries are the eigenvalues λ_k of H. Thus,

$$SL^{-1}AL^{-T}S^{-1} = E,$$
$$A^{-1}B = L^{-T}S^{-1}E^{-1}SL^{-1}(LL^T) = P^{-1}E^{-1}P, \qquad (4.11.2)$$

where $P \equiv SL^T$.

From 4.11.1 it is easy to see that

$$\mathbf{u}_n = (A^{-1}B)^n \mathbf{u}_0,$$

so that, using 4.11.2

$$\mathbf{u}_n = (P^{-1}E^{-1}P)^n \mathbf{u}_0,$$
$$\mathbf{u}_n = P^{-1}E^{-n}P\mathbf{u}_0 \qquad (4.11.3)$$

and

$$\lambda_1^n \mathbf{u}_n = P^{-1}[\lambda_1^n E^{-n}]P\mathbf{u}_0.$$

The diagonal matrix in square brackets has diagonal entries of the form $(\lambda_1/\lambda_k)^n$. Since λ_1 is assumed to be the smallest eigenvalue in absolute value, as $n \to \infty$ each diagonal term converges to either 0 (if $|\lambda_k| > |\lambda_1|$) or 1 (if $\lambda_k = \lambda_1$). Thus the matrix in square brackets converges to some constant diagonal matrix, F, which has $F_{k,k} = 1$ when $E_{k,k} = \lambda_1$, and $F_{k,k} = 0$ otherwise. For future reference, note that this implies

$$EF = \lambda_1 F. \qquad (4.11.4)$$

Therefore,

$$\lim_{n \to \infty} \lambda_1^n \mathbf{u}_n = P^{-1}FP\mathbf{u}_0. \qquad (4.11.5)$$

Unless the vector $P^{-1}FP\mathbf{u_0}$ is equal to the zero vector, an event that has probability zero if $\mathbf{u_0}$ is chosen "randomly," then it is an eigenvector of $A\mathbf{u} = \lambda B\mathbf{u}$. For, using 4.11.2 and 4.11.4

$$A(P^{-1}FP\mathbf{u_0}) = (BP^{-1}EP)(P^{-1}FP\mathbf{u_0}) = \lambda_1 B(P^{-1}FP\mathbf{u_0}).$$

Thus, according to 4.11.5, for large n, $\mathbf{u_n}$ is approximately equal to an (unnormalized) eigenvector corresponding to the smallest eigenvalue of the eigenproblem $A\mathbf{u} = \lambda B\mathbf{u}$.

Once an eigenvector approximation $\mathbf{u_{n+1}}$ is available, it is trivial to approximate the corresponding eigenvalue. Since $B\mathbf{u_n} = A\mathbf{u_{n+1}} \approx \lambda_1 B\mathbf{u_{n+1}}$, $\mathbf{u_n} \approx \lambda_1 \mathbf{u_{n+1}}$, all we have to do to estimate λ_1 is to take the ratio of the components of $\mathbf{u_n}$ to those of $\mathbf{u_{n+1}}$.

However, μ_n in 4.11.1 is a particularly accurate way to approximate λ_1. For, from 4.11.3,

$$\mu_n = \frac{\mathbf{u_n}^T B \mathbf{u_n}}{\mathbf{u_n}^T B \mathbf{u_{n+1}}} = \frac{[P^{-1}E^{-n}P\mathbf{u_0}]^T B [P^{-1}E^{-n}P\mathbf{u_0}]}{[P^{-1}E^{-n}P\mathbf{u_0}]^T B [P^{-1}E^{-(n+1)}P\mathbf{u_0}]}$$

$$= \frac{\mathbf{u_0}^T P^T E^{-n}[P^{-T}BP^{-1}]E^{-n}P\mathbf{u_0}}{\mathbf{u_0}^T P^T E^{-n}[P^{-T}BP^{-1}]E^{-(n+1)}P\mathbf{u_0}}.$$

By using the relations $P = SL^T$, $B = LL^T$, and $S^T = S^{-1}$, it is easy to verify that $P^{-T}BP^{-1} = I$, and so

$$\mu_n = \lambda_1 \frac{\mathbf{u_0}^T P^T [\lambda_1^{2n} E^{-2n}] P\mathbf{u_0}}{\mathbf{u_0}^T P^T [\lambda_1^{2n+1} E^{-(2n+1)}] P\mathbf{u_0}}.$$

The two terms in square brackets are diagonal matrices with entries $(\lambda_1/\lambda_k)^{2n}$ and $(\lambda_1/\lambda_k)^{2n+1}$, respectively. Thus, as discussed above, these matrices both converge to F, and μ_n converges to λ_1 as $n \to \infty$. However, notice that since the rate at which these matrices converge to F is determined by the ratio $|\lambda_1/\lambda_2|$, where λ_2 is the *second*-smallest eigenvalue in absolute value, the convergence rate is now twice as fast as before, because the exponent of this ratio is $2n$ (or $2n + 1$), rather than n. ∎

The iteration 4.11.1 can be easily modified to calculate eigenvalues other than the smallest one. In the shifted inverse power iteration,

$$(A - \alpha B)\mathbf{u_{n+1}} = B\mathbf{u_n},$$

$$\mu_n = \alpha + \frac{\mathbf{u_n}^T B \mathbf{u_n}}{\mathbf{u_n}^T B \mathbf{u_{n+1}}}, \tag{4.11.6}$$

μ_n will converge to the eigenvalue of $A\mathbf{u} = \lambda B\mathbf{u}$ closest to α, and $\mathbf{u_n}$ will converge to a corresponding (unnormalized) eigenvector. From Theorem 4.11.1, we know that for large n, $\mathbf{u_n}$ will be an eigenvector of the shifted problem $(A - \alpha B)\mathbf{u} = \lambda B\mathbf{u}$ corresponding to the smallest eigenvalue (in absolute value) of that problem, and that $\mu_n - \alpha$ will approximate that smallest eigenvalue. So, for large n,

$$(A - \alpha B)\mathbf{u_n} \approx (\mu_n - \alpha)B\mathbf{u_n},$$

$$A\mathbf{u_n} \approx \mu_n B\mathbf{u_n}.$$

Therefore (for large n), μ_n is an eigenvalue of $A\mathbf{u} = \lambda B\mathbf{u}$, and, since $\mu_n - \alpha$ is the smallest of all the eigenvalues of the shifted problem, μ_n is the eigenvalue of $A\mathbf{u} = \lambda B\mathbf{u}$ closest to α.

The iteration 4.11.6 can be used not only to find eigenvalues of $A\mathbf{u} = \lambda B\mathbf{u}$ other than the smallest, but it can also be used to accelerate the inverse power method convergence. The convergence rate, as noted earlier, depends on the ratio of the smallest eigenvalue to the second smallest. For the shifted problem, this ratio is

$$\frac{|\lambda_{\text{closest}} - \alpha|}{|\lambda_{\text{next}} - \alpha|},$$

where λ_{closest} and λ_{next} are the eigenvalues of the original problem ($A\mathbf{u} = \lambda B\mathbf{u}$) closest and next closest, respectively, to α. Thus, clearly, the closer α is chosen to the desired eigenvalue, the more rapid the convergence of 4.11.6 will be.

To illustrate the use of the inverse power method, consider the problem

$$-u_{xx} + 2u_x - u = \lambda u,$$

$$u(0) = 0, \qquad\qquad\qquad\qquad (4.11.7)$$

$$u_x(1) = 0.$$

The first two eigenvalues are $\lambda_1 = 4.1158584$, $\lambda_2 = 24.1393420$, and the eigenfunctions are $e^x \sin(\sqrt{\lambda_i}x)$. Let us discretize this problem in a straightforward, *nonsymmetric*, manner:

$$\frac{-U(x_{i+1}) + 2U(x_i) - U(x_{i-1})}{\Delta x^2} + \frac{U(x_{i+1}) - U(x_{i-1})}{\Delta x}$$

$$- U(x_i) = \lambda U(x_i) \qquad i = 1, \ldots, N,$$

$$U(x_0) = 0, \qquad\qquad\qquad\qquad (4.11.8)$$

$$\frac{U(x_{N+1}) - U(x_{N-1})}{2\,\Delta x} = 0.$$

If $U(x_{N+1})$ in the Nth equation above is replaced by $U(x_{N-1})$ (this follows from the second boundary condition approximation), then equations 4.11.8 represent a matrix eigenvalue problem of the form $A\mathbf{u} = \lambda B\mathbf{u}$ where $B = I$, $\mathbf{u}$ is the vector $(U(x_1), \ldots, U(x_N))$, and A is a *nonsymmetric* tridiagonal matrix with elements

$$A_{i,i-1} = \frac{-1}{\Delta x^2} - \frac{1}{\Delta x} \qquad \text{except } A_{N,N-1} = \frac{-2}{\Delta x^2},$$

$$A_{i,i} = \frac{2}{\Delta x^2} - 1,$$

$$A_{i,i+1} = \frac{-1}{\Delta x^2} + \frac{1}{\Delta x}.$$

The shifted inverse power iteration 4.11.6 was used to estimate the smallest eigenvalues of this discrete problem, using various values for Δx and α. The tridiagonal solver of Figure 0.4.6 was used to calculate $\mathbf{u}_{n+1}$ each step of the iteration, and a starting vector of $\mathbf{u}_0 = (1, 1, \ldots, 1)$ was used in each case. The results are summarized in Table 4.11.1. The "maximum error in eigenfunction" is calculated after normalizing the approximate and exact eigenfunctions, and the "number of iterations" is the number required to produce eight significant figures in the eigenvalue.

As expected, the error in both the eigenvalues and the eigenfunctions appears to be $O(\Delta x^2)$. As was also expected, choosing α close to the desired root accelerated the convergence substantially. Notice that the fact that A is nonsymmetric did not prevent the inverse power method from converging. In fact, it can be shown that, while symmetry makes the theoretical development much easier, the only real requirement for convergence of the inverse power

Table 4.11.1
Inverse Power Method Results

Δx	α	Number of iterations	μ_{final}	Error in eigenvalue	Maximum error in eigenfunction
0.02	0.0	11	4.1138323	0.0020261	0.0001286
	4.0	4	4.1138323	0.0020261	0.0001286
	20.0	16	24.1092630	0.0300790	0.0003775
0.01	0.0	11	4.1153518	0.0005066	0.0000322
	4.0	4	4.1153518	0.0005066	0.0000322
	20.0	16	24.1318191	0.0075229	0.0000943

method is that there must be a smallest eigenvalue. This rules out complex eigenvalues, since they occur in conjugate pairs with equal absolute values.

4.12 Problems

1. Show that the steady-state problem

$$0 = \mathbf{V}^T(D(x, y, z)\mathbf{V}\mathbf{u})$$
$$+ a(x, y, z)u + f(x, y, z) \qquad \text{in } R,$$
$$u = g_0(x, y, z) \qquad \text{on part of the boundary,}$$
$$\partial u/\partial n = g_1(x, y, z) \qquad \text{on the other part,}$$

has at most one (smooth) solution, if $D > 0$, $a \leq 0$, and the part of the boundary where u is given is nonempty.

Hint: Assume u_1 and u_2 are two solutions, multiply the partial differential equation satisfied by e by $e \equiv u_1 - u_2$, and integrate over R, to show that $\mathbf{V}e \equiv \mathbf{0}$.

2. a. Verify that the potential energy function given in 4.0.2 satisfies Laplace's equation.
   b. Verify that 4.0.4 is a solution of the ordinary differential equation 4.0.3.

3. Consider the boundary value problem (exact solution $= \sin(2x)$):

$$-u_{xx} + u = 5\sin(2x),$$
$$u(0) = u(\pi),$$
$$u_x(0) = u_x(\pi).$$

These are called "periodic" boundary conditions. If $x_i = i\,\Delta x$ and $\Delta x = \pi/N$, the boundary conditions may be approximated by

$$U(x_0) = U(x_N),$$
$$\frac{U(x_1) - U(x_0)}{\Delta x} = \frac{U(x_{N+1}) - U(x_N)}{\Delta x} \qquad \text{or} \qquad U(x_{N+1}) = U(x_1).$$

If at x_i, $i = 1, \ldots, N$, u_{xx} is replaced by its usual centered difference approximation, and the boundary approximations are used to eliminate $U(x_0)$ and $U(x_{N+1})$ from the first and last equations, there results a system of linear equations whose coefficient matrix has the nonzero structure

$$
\begin{bmatrix}
X & X & 0 & 0 & 0 & 0 & 0 & & 0 & 0 & X \\
X & X & X & 0 & 0 & 0 & 0 & & 0 & 0 & 0 \\
0 & X & X & X & 0 & 0 & 0 & & 0 & 0 & 0 \\
0 & 0 & X & X & X & 0 & 0 & \cdots & 0 & 0 & 0 \\
0 & 0 & 0 & X & X & X & 0 & & 0 & 0 & 0 \\
0 & 0 & 0 & 0 & X & X & X & & 0 & 0 & 0 \\
0 & 0 & 0 & 0 & 0 & X & X & & 0 & 0 & 0 \\
 & & \vdots & & & & & & & \vdots & \\
0 & 0 & 0 & 0 & 0 & 0 & 0 & \cdots & X & X & 0 \\
0 & 0 & 0 & 0 & 0 & 0 & 0 & & X & X & X \\
X & 0 & 0 & 0 & 0 & 0 & 0 & & 0 & X & X
\end{bmatrix}.
$$

Write a FORTRAN subroutine which requires only $O(N)$ work and storage to solve a linear system with a coefficient matrix of this structure (you may assume no pivoting is necessary). Then use this subroutine to solve the finite difference system, with $N = 20$.

4. Show that the finite difference approximation 4.1.2 to 4.1.1 is stable, provided $a(x_i) \leq -\rho < 0$. (This assumption can be relaxed, but more work is required.) In other words, show that when the truncation error goes to zero, the error does also.

   Hint: Show that the error e satisfies an equation like 4.1.3, but with $f(x_i)$ replaced by the trucation error, T_i. Then note that this equation is similar to 2.2.3 (with $\Delta t = \infty$!). Use an argument similar to that used to prove stability for 2.2.1 to prove that $E \leq T_{\max}/\rho$ when Δx is sufficiently small, where $E \equiv \max |e(x_i)|$ and $T_{\max} \equiv \max |T_i|$. Note that what you are showing is that the tridiagonal linear systems associated with 4.1.3 have inverses that are bounded as $\Delta x \to 0$.

5. Show that it is impossible to approximate $u_{xx}(x_i)$ by a formula of the form

$$
Au(x_{i+1}) + Bu(x_i) + Cu(x_{i-1}),
$$

   which has $O(\Delta x_{\max}^2)$ accuracy, unless the points x_i are uniformly spaced.

6. Consider the rigid elastic beam problem 4.9.1, solved previously by reformulation as a system of two second-order equations.

a. Show that the finite difference equation ($x_i = i\,\Delta x$, $\Delta x = 1/N$)

$$\frac{U(x_{i+2}) - 4U(x_{i+1}) + 6U(x_i) - 4U(x_{i-1}) + U(x_{i-2})}{\Delta x^4} = 0$$

$$\text{for } i = 1, \ldots, N-1,$$

is consistent with the differential equation 4.9.1 with truncation error $O(\Delta x^2)$.

b. Using *second-order* accurate formulas to approximate the boundary conditions in 4.9.1, eliminate $U(x_{-1})$, $U(x_0)$, $U(x_N)$, and $U(x_{N+1})$ from the $N-1$ finite difference equations above.

c. The resulting finite difference equations form a banded linear system with "half-bandwidth" $L = 2$. Solve this system, with $\Delta x = 0.02$ and 0.01, by using the band solver of Figure 0.4.3. Compare your results with the differential equation solution $u(x) = -0.5x^3 + 1.5x^2$, and estimate the experimental order of convergence.

7. The fourth-order boundary value problem 4.9.1 can also be solved by using a shooting method. First, let us reformulate it as four first-order equations:

$$u_1' = u_2, \qquad u_1(0) = 0, \qquad u_1(1) = 1,$$
$$u_2' = u_3, \qquad u_2(0) = 0, \qquad u_3(1) = 0,$$
$$u_3' = u_4,$$
$$u_4' = 0.$$

The shooting problem is to determine parameters α and β such that when this ordinary differential equation system is solved with initial conditions

$$u_1(0) = 0, \qquad u_3(0) = \alpha,$$
$$u_2(0) = 0, \qquad u_4(0) = \beta,$$

the functions

$$f_1(\alpha, \beta) \equiv u_1(1) - 1,$$
$$f_2(\alpha, \beta) \equiv u_3(1)$$

are zero. Use Newton's method to solve this system of two equations, calculating the Jacobian elements by using finite differences:

$$\frac{\partial f_i}{\partial \alpha}(\alpha_n, \beta_n) \approx \frac{f_i(\alpha_n + \Delta\alpha, \beta_n) - f_i(\alpha_n, \beta_n)}{\Delta\alpha}$$

$$\frac{\partial f_i}{\partial \beta}(\alpha_n, \beta_n) \approx \frac{f_i(\alpha_n, \beta_n + \Delta\beta) - f_i(\alpha_n, \beta_n)}{\Delta\beta}$$

$$i = 1, 2.$$

Since the differential equations are linear, the functions f_1 and f_2 are linear, and the Jacobian element approximations given above are exact (to within roundoff) for any nonzero $\Delta\alpha$, $\Delta\beta$. Also, because of the linearity of f_1 and f_2, only one iteration of Newton's method is needed, using any starting values. Thus, take $\alpha_0 = \beta_0 = 0$ and $\Delta\alpha = \Delta\beta = 1$, and solve the initial value problem three times, with (α, β) equal to $(0, 0)$, $(1, 0)$, and $(0, 1)$, by using any good initial value solver (e.g., IMSL's IVPRK) in your computer library, with a small error tolerance. Do one iteration of Newton's method to calculate the correct values of α and β. (Answer: $\alpha = 3$ and $\beta = -3$.)

8. Show that when the partial differential equation

$$0 = D(x, y, z) \nabla^2 u + a(x, y, z)u$$
$$+ f(x, y, z) \qquad \text{in } (0, 1) \times (0, 1) \times (0, 1),$$
$$u = g(x, y, z) \qquad \text{on the boundary}$$

where $D > 0$, $a \leq 0$, is approximated by a centered finite difference formula (cf. 4.5.1), the resulting linear system has a unique solution. Hint: Follow the proof that 4.1.2 has a unique solution.

9. Solve the finite difference equations associated with the three-dimensional problem 4.7.6 using successive over-relaxation, with $N = 10$ and $N = 20$ ($N = 1/\Delta x = 1/\Delta y = 1/\Delta z$) and with $\omega = 1.00, 1.05, \ldots, 1.95$ for each of the two stepsizes. Count the number of iterations to convergence (use zero initial values and stop when the maximum change per iteration in the solution is less than some fixed tolerance) and pick out the optimal ω for each value of N. Verify that the number of iterations when $\omega = \omega_{\text{opt}}$ is $O(N)$, and that the total work is $O(N^4)$.

10. a. Show that the general second-order linear boundary value problem in one variable 4.1.1 can be put into the "symmetric" form

$$[p(x)u_x]_x + q(x)u = g(x)$$

by multiplying by an appropriate function.

b. Approximate this by a finite difference formula similar to 4.10.3, which has $O(\Delta x^2)$ truncation error and generates a *symmetric* tridiagonal linear system. If boundary conditions more general than those in 4.1.1 are used, how can this linear system be modified to make it symmetric, if it is not already?

Hint: Note that the boundary conditions affect only the first and last equations.

c. Show that the linear partial differential equation

$$u_{xx} + u_{yy} + a(x, y)u_x + b(x, y)u_y + c(x, y)u = f(x, y)$$

can be put into the form

$$[A(x, y)u_x]_x + [A(x, y)u_y]_y + C(x, y)u = F(x, y),$$

provided $a_y = b_x$.

d. Approximate this by a second-order finite difference formula that generates a symmetric linear system.

11. a. Show that with the change of variables

$$\mathbf{x_n} = L^T\mathbf{X_n},$$

$$\mathbf{r_n} = L^T\mathbf{R_n},$$

$$\mathbf{p_n} = L^T\mathbf{P_n},$$

the "preconditioned" conjugate-gradient method,

$$\mathbf{X_0} = \text{starting guess at solution},$$

$$\mathbf{R_0} = B^{-1}(\mathbf{b} - A\mathbf{X_0}),$$

$$\mathbf{P_0} = \mathbf{R_0},$$

$n = 1, \dots$

$$\lambda_n = (\mathbf{R_n^T}B\mathbf{P_n})/(\mathbf{P_n^T}A\mathbf{P_n}),$$

$$\mathbf{X_{n+1}} = \mathbf{X_n} + \lambda_n\mathbf{P_n},$$

$$\mathbf{R_{n+1}} = \mathbf{R_n} - \lambda_n B^{-1}A\mathbf{P_n},$$

$$\alpha_n = -(\mathbf{R_{n+1}^T}A\mathbf{P_n})/(\mathbf{P_n^T}A\mathbf{P_n}),$$

$$\mathbf{P_{n+1}} = \mathbf{R_{n+1}} + \alpha_n\mathbf{P_n}$$

reduces to the usual conjugate-gradient method 4.8.1 applied to the linear system $H\mathbf{x} = \mathbf{c}$, where $H = L^{-1}AL^{-T}$, $\mathbf{c} = L^{-1}\mathbf{b}$. Assume that both A and B are positive-definite, and that $B = LL^T$ is the Cholesky decomposition of B (see Theorem 0.3.1).

b. Verify that H is positive-definite, and therefore that the conjugate-gradient method applied to $Hx = c$ is guaranteed to converge (in the absence of roundoff error). Show that this assures that the preconditioned conjugate-gradient iterate X_n converges to the solution of the original problem $AX = b$.

c. Show that H and $B^{-1}A$ have the same eigenvalues and therefore the same spectral condition number.

   If B is chosen so that $B^{-1}A$ has a smaller condition number than A, the preconditioned conjugate-gradient method above should converge faster than the unmodified version. B is generally chosen to be some approximation to A such that $Bz = d$ is easy to solve, since B^{-1} appears in the iteration. The simplest choice is $B = \text{diagonal}(A)$.

12. Use the inverse power method 4.11.1 to find the smallest eigenvalue of the matrix eigenvalue problem 4.10.7, which approximates the membrane eigenvalue problem 4.10.6. You may use either successive over-relaxation or the band solver of Figure 0.4.3 to solve the linear system 4.11.1 each step. Use $\Delta x = \Delta y = 0.1$ so that you can compare your result with that given in Table 4.10.1, namely $\lambda_1 = 19.59$.

5

The Finite Element Method

5.0 Introduction

The finite element method was first used by engineers to solve structural problems. They modeled a continuous structure using a number of "finite" (distinct from infinitesimally small) elements that were connected at certain nodal points, and they required the forces to balance at each node. Only later was it realized that this technique could be thought of as a numerical method for solving the partial differential equations modeling the stresses in a continuous structure, and that similar methods could be used to solve other differential equations.

Now the "finite element method" (really a class of methods) is generally considered to be a competitor of the finite difference methods and is used to solve as wide a range of ordinary and partial differential equations as the latter. Although finite element methods are usually substantially more difficult to program, this extra effort yields approximations that are of high-order accuracy even when a partial differential equation is solved in a general (nonrectangular) multi-dimensional region, and even when the solution varies more rapidly in certain portions of the region so that a uniform grid is not appropriate. These and other considerations have earned the finite element method great popularity in recent years both for initial value and (especially) boundary value differential equations.

5.1   The Galerkin Method for Boundary Value Problems

The most widely used form of the finite element method is the "Galerkin" method. Although other forms are also in use (an example using the "collocation" version will be given in Section 5.4) we shall concentrate our attention on the Galerkin finite element method.

Although the Galerkin method can be applied to much more general problems, the following three-dimensional linear boundary value problem (cf. the diffusion problem 4.0.1) is chosen to make the analysis simple. Its use on more general problems will be illustrated by an example in Section 5.3. We study the problem

$$0 = \mathbf{V}^T(D(x, y, z)\mathbf{V}u) - a(x, y, z)u + f(x, y, z) \qquad \text{in } R,$$

$$u = r(x, y, z) \qquad \text{on part of the boundary of } R \ (\partial R_1), \qquad (5.1.1)$$

$$D\partial u/\partial n = -p(x, y, z)u + q(x, y, z) \qquad \text{on the other part } (\partial R_2).$$

It is assumed that $D > 0$.

Here $\partial u/\partial n$ represents the directional derivative of u in the direction of the unit outward normal to the boundary. Note that $D\partial u/\partial n = D\mathbf{V}u^T\mathbf{n} = -\mathbf{J}^T\mathbf{n}$ is the inward boundary flux of the diffusing component.

If the partial differential equation and second boundary condition are multiplied by a smooth function Φ that is arbitrary except that it is required to satisfy $\Phi = 0$ on ∂R_1, and if these are integrated over R and ∂R_2 respectively, then

$$0 = \iiint_R \left[\mathbf{V}^T(D\mathbf{V}u) - au + f \right]\Phi \, dx \, dy \, dz + \iint_{\partial R_2} \left[-D\frac{\partial u}{\partial n} - pu + q \right]\Phi \, dA.$$

Using $\Phi\mathbf{V}^T(D\mathbf{V}u) = \mathbf{V}^T(\Phi D\mathbf{V}u) - D\mathbf{V}u^T\mathbf{V}\Phi$ and the divergence theorem:

$$0 = \iiint_R \left[-D\mathbf{V}u^T\mathbf{V}\Phi - au\Phi + f\Phi \right] dx \, dy \, dz + \iint_{\partial R} \Phi D\mathbf{V}u^T\mathbf{n} \, dA$$

$$+ \iint_{\partial R_2} \left[-\Phi D\mathbf{V}u^T\mathbf{n} - pu\Phi + q\Phi \right] dA.$$

The integrand in the first boundary integral is nonzero only on ∂R_2, since $\Phi = 0$ on ∂R_1. Thus

$$0 = \iiint\limits_R \left[-D\nabla \mathbf{u}^T \nabla \Phi - au\Phi + f\Phi \right] dx\, dy\, dz + \iint\limits_{\partial R_2} \left[-pu\Phi + q\Phi \right] dA.$$

$$(5.1.2)$$

Equation 5.1.2 is called the weak formulation of the partial differential equation 5.1.1. It is almost equivalent to 5.1.1 in the sense that if u is *smooth* and satisfies 5.1.2 for any smooth Φ vanishing on ∂R_1, the steps leading from 5.1.1 to 5.1.2 can be reversed, so that u satisfies the partial differential equation and the second (natural) boundary condition. As part of either formulation. u is required to satisfy the first (essential) boundary condition.

The Galerkin method attempts to find an approximate solution to the weak formulation 5.1.2 of the form

$$U(x, y, z) = \Omega(x, y, z) + \sum_{i=1}^{M} a_i \Phi_i(x, y, z), \qquad (5.1.3)$$

where $\{\Phi_1, \ldots, \Phi_M\}$ is a set of linearly independent "trial" functions that vanish on ∂R_1 and Ω is another function that satisfies the essential boundary condition $\Omega = r$ on ∂R_1. Clearly, U will satisfy $U = r$ on ∂R_1 regardless of the values chosen for $a_1 \ldots a_M$.

It is impossible to find parameters a_i such that U satisfies 5.1.2 for arbitrary Φ vanishing on ∂R_1, since we only have a finite number of parameters. Thus, it is only required that 5.1.2 be satisfied for $\Phi = \Phi_1, \ldots, \Phi_M$ (each of which vanishes on ∂R_1):

$$0 = \iiint\limits_R \left[-D\nabla U^T \nabla \Phi_k - aU\Phi_k + f\Phi_k \right] dx\, dy\, dz$$

$$+ \iint\limits_{\partial R_2} \left[-pU\Phi_k + q\Phi_k \right] dA. \qquad (5.1.4)$$

This can be written as a system of M linear equations for the M unknown parameters $a_1, \ldots, a_M$:

$$\sum_{i=1}^{M} A_{ki} a_i = b_k,$$

where

$$b_k = \iiint\limits_{R} \left[f\Phi_k - D\mathbf{V}\Omega^T\mathbf{V}\Phi_k - a\Omega\Phi_k \right] dx\,dy\,dz + \iint\limits_{\partial R_2} \left[q\Phi_k - p\Omega\Phi_k \right] dA,$$

(5.1.5a)

$$A_{ki} = \iiint\limits_{R} \left[D\mathbf{V}\Phi_k^T\mathbf{V}\Phi_i + a\Phi_k\Phi_i \right] dx\,dy\,dz + \iint\limits_{\partial R_2} \left[p\Phi_k\Phi_i \right] dA.$$

(5.1.5b)

The finite element choices for $\Omega, \Phi_1, \ldots, \Phi_M$ are piecewise polynomial functions that are each zero outside some small subregion of R. This ensures that A is sparse, for $A_{ki} = 0$ (see 5.1.5b) unless the regions where Φ_k and Φ_i are nonzero overlap.

From 5.1.5b it is clear that the matrix A is symmetric. If, in addition to the assumption $D > 0$, it is also assumed that $a \geq 0$, $p \geq 0$, and that ∂R_1 is nonempty, then the matrix A is also positive-definite. To verify this, let $\mathbf{z}$ be an arbitrary vector with components $z_1, \ldots, z_M$. Then

$$\mathbf{z}^T A \mathbf{z} = \sum_{k=1}^{M} \left(\sum_{i=1}^{M} A_{ki} z_i \right) z_k$$

$$= \iiint\limits_{R} \left[D|\mathbf{V}Z|^2 + aZ^2 \right] dx\,dy\,dz + \iint\limits_{\partial R_2} pZ^2\,dA \geq 0,$$

where $Z(x, y, z) \equiv \sum z_i \Phi_i(x, y, z)$. If equality holds ($\mathbf{z}^T A \mathbf{z} = 0$), then clearly (since $D > 0$), $\mathbf{V}Z \equiv \mathbf{0}$, so that $Z(x, y, z)$ is a constant. Since it is assumed that ∂R_1 is not empty, the fact that $Z = 0$ on ∂R_1 implies it must be zero everywhere, and, by the independence of the Φ_i, the z_i must all be zero.

This completes the proof that A is positive-definite. Not only will the linear system $A\mathbf{a} = \mathbf{b}$ have a unique solution, but Gaussian elimination without pivoting is guaranteed to work when this system is solved, by Theorem 0.2.4. In addition, the successive over-relaxation and conjugate-gradient iterative methods discussed in Chapter 4, as well as most other popular iterative methods, are guaranteed to converge when applied to $A\mathbf{a} = \mathbf{b}$.

To obtain a bound on the Galerkin method error, we assume as above that $D > 0$, $a \geq 0$, $p \geq 0$ and $\partial R_1 \neq \varnothing$, and we introduce an "energy" functional defined by

$$F(u) \equiv \iiint\limits_{R} \{\tfrac{1}{2}D|\mathbf{V}u|^2 + \tfrac{1}{2}au^2 - fu\}\,dx\,dy\,dz + \iint\limits_{\partial R_2} \{\tfrac{1}{2}pu^2 - qu\}\,dA.$$

If $w(x, y, z)$ is any smooth function satisfying the essential boundary condition on ∂R_1, and $e \equiv w - u$, where u is the partial differential equation solution, then

$$F(w) = F(u + e) = F(u)$$

$$+ \left[\iiint\limits_{R} \{D\nabla u^T \nabla e + aue - fe\} \, dx \, dy \, dz + \iint\limits_{\partial R_2} \{pue - qe\} \, dA \right]$$

$$+ \iiint\limits_{R} \{\tfrac{1}{2}D|\nabla e|^2 + \tfrac{1}{2}ae^2\} \, dx \, dy \, dz + \iint\limits_{\partial R_2} \tfrac{1}{2}pe^2 \, dA.$$

The term in square brackets is zero, since u satisfies the weak formulation 5.1.2 and e is a function that vanishes on ∂R_1. Thus,

$$F(w) = F(u) + \|w - u\|_F^2, \tag{5.1.6}$$

where the "energy norm" is defined by

$$\|e\|_F^2 \equiv \iiint\limits_{R} \{\tfrac{1}{2}D|\nabla e|^2 + \tfrac{1}{2}ae^2\} \, dx \, dy \, dz + \iint\limits_{\partial R_2} \tfrac{1}{2}pe^2 \, dA.$$

The assumptions made on D, a, p, and ∂R_1 ensure that this is a legitimate norm and that it is equal to zero only if $e \equiv 0$.

Among all functions satisfying the essential boundary condition ($w = r$ on ∂R_1), therefore, the solution u of the partial differential equation minimizes the energy functional.

We can also show that U minimizes the energy fuctional among all trial functions of the form 5.1.3. For if W is any such function, and $E \equiv W - U$, then

$$F(W) = F(U + E) = F(U)$$

$$+ \left[\iiint\limits_{R} \{D\nabla U^T \nabla E + aUE - fE\} \, dx \, dy \, dz + \iint\limits_{\partial R_2} \{pUE - qE\} \, dA \right]$$

$$+ \|E\|_F^2.$$

Since $E(x, y, z)$ is a linear combination of the $\Phi_1, \ldots, \Phi_M$, 5.1.4 ensures that the term in square brackets is again zero. So

$$F(W) = F(U) + \| W - U \|_F^2.$$

and U minimizes F over the set of trial functions.
Thus

$$F(U) \leq F(W),$$

$$F(U) - F(u) \leq F(W) - F(u).$$

Since both U and W satisfy the essential boundary condition, 5.1.6 can be applied with either $w = U$ or $w = W$, giving

$$\| U - u \|_F \leq \| W - u \|_F. \tag{5.1.7}$$

This result is the basis on which the Galerkin error can be estimated, since 5.1.7 states that the finite element approximation U is as close to the true solution u as any other function W in the trial set 5.1.3, *in the energy norm*. The energy norm is not a very convenient measure of error, however, so in practice it is generally assumed that 5.1.7 holds, or almost holds, for any convenient norm, and, in fact, theoretical justification for this assumption exists for the more widely used norms. That is, in practice the question of the accuracy of the Galerkin method is reduced to the question: how accurately can I approximate the true solution using *any* function in the trial space 5.1.3?

In any case, if the distance from the solution to the trial space goes to zero in the energy norm as $M \to \infty$, 5.1.7 guarantees convergence of the finite element solution.

5.2 An Example Using Piecewise Linear Trial Functions

As an example of the use of the Galerkin finite element method, consider the problem

$$0 = u_{xx} - a(x)u + f(x),$$

$$u(0) = r, \tag{5.2.1}$$

$$u_x(1) = -pu(1) + q.$$

This is the one-dimensional version of 5.1.1, with $D = 1$, $\partial R_1 = \{x = 0\}$, and $\partial R_2 = \{x = 1\}$.

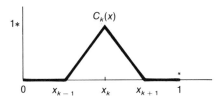

Figure 5.2.1

The basis functions in the expansion 5.1.3 for $U(x)$ are chosen from the "chapeau" functions (see Figure 5.2.1):

$$
C_k(x) =
\begin{cases}
\dfrac{x - x_{k-1}}{x_k - x_{k-1}}, & \text{for } x_{k-1} \le x \le x_k, \\[2mm]
\dfrac{x_{k+1} - x}{x_{k+1} - x_k}, & \text{for } x_k < x \le x_{k+1}, \\[2mm]
0, & \text{elsewhere}
\end{cases}
$$

where $x_{-1} < 0 = x_0 < x_1 < \cdots < x_N = 1 < x_{N+1}$. Notice that $C_k(x_i) = 0$ unless $i = k$, in which case it is 1. We choose $\Omega(x) = rC_0(x)$ and $\Phi_k(x) = C_k(x)$ $(k = 1, \ldots, N)$ so that $\Omega(0) = r$, while $\Phi_k(0)$ is zero for each $k = 1, \ldots, N$. Thus $U(x)$ will satisfy the essential boundary condition at zero, regardless of the values of the parameters in the expansion 5.1.3.

In this example, the nodes will be taken to be $x_k = k\Delta x$ ($\Delta x = 1/N$), but notice that it is not necessary to partition the interval uniformly.

The formulas (5.1.5a–b) may be used to calculate the matrix and right-hand side elements for problem 5.2.1, provided the integrals over R are replaced by integrals over the interval $(0, 1)$ and integrals over ∂R_2 are replaced by evaluation at $x = 1$. Notice also that at $x = 1$, $D\partial u/\partial n = u_x(1)$, so that the natural boundary condition in 5.2.1 is in the form required by 5.1.1. The number of unknowns is now $M = N$.

This gives

$$
A_{ki} = \int_0^1 \left[\Phi_k' \Phi_i' + a\Phi_k \Phi_i \right] dx + p\Phi_k(1)\Phi_i(1),
$$

$$
b_k = \int_0^1 \left[f\Phi_k - rC_0' \Phi_k' - arC_0 \Phi_k \right] dx + q\Phi_k(1) - prC_0(1)\Phi_k(1).
$$

Since Φ_k is nonzero only in the subinterval $[x_{k-1}, x_{k+1}]$, A is clearly tridiagonal. The above integrals cannot be evaluated in closed form for arbitrary $a(x)$ and $f(x)$, so the trapezoid rule is used to approximate them numerically. That is, the integral of each integrand $g(x)$ is approximated by

$$\int_0^1 g(x)\, dx \approx \sum_{i=1}^{N} (x_i - x_{i-1})[\tfrac{1}{2}g(x_{i-1}) + \tfrac{1}{2}g(x_i)].$$

The use of this numerical integration rule will be justified in Section 5.3. After considerable effort, this yields

$$A_{k,k} = \frac{2}{\Delta x} + a(x_k)\Delta x, \qquad \text{(except)}\ A_{N,N} = \frac{1}{\Delta x} + \tfrac{1}{2}a(x_N)\Delta x + p,$$

$$A_{k,k+1} = A_{k+1,k} = \frac{-1}{\Delta x}, \tag{5.2.2}$$

$$b_k = f(x_k)\Delta x, \qquad \text{(except)}\ b_1 = f(x_1)\Delta x + \frac{r}{\Delta x},$$

$$b_N = \tfrac{1}{2}f(x_N)\Delta x + q.$$

A closer look shows that these equations are precisely the same as the usual second-order finite difference equations! To see this, first notice that, since

$$U(x) = rC_0(x) + \sum a_i \Phi_i(x),$$

we have $U(x_k) = a_k$ ($k = 1, \ldots, N$). Thus the (tridiagonal) finite element equations can be written

$$A_{k,k-1}U(x_{k-1}) + A_{k,k}U(x_k) + A_{k,k+1}U(x_{k+1}) = b_k$$

or, using formulas 5.2.2 and simplifying,

$$\frac{U(x_{k+1}) - 2U(x_k) + U(x_{k-1})}{\Delta x^2} - a(x_k)U(x_k) + f(x_k) = 0, \tag{5.2.3}$$

when $2 \le k \le N - 1$. When $k = 1$ and $k = N$, the equations reduce to the above finite difference formula, with

$$U(x_0) = r$$

and

$$U(x_{N+1}) = U(x_{N-1}) + 2\Delta x[-pU(x_N) + q],$$

respectively. The latter formula is a central difference approximation to the natural boundary condition (at $x = 1$) in the partial differential equation problem 5.2.1. The formula 5.2.3 is easily recognized as the usual central finite difference approximation to 5.2.1 (cf. 4.1.2 with $D = 1$, $v = 0$).

Since it is equivalent to a centered difference method, we know that the finite element method used above will have $O(\Delta x^2)$ error, but nevertheless, let us verify this error estimate using 5.1.7. Let us see how closely we can approximate the true solution u by using a function W in the trial space 5.1.3. We take as our W the piecewise linear function that interpolates to the true solution at $x_0, \ldots, x_N$:

$$W(x) = \sum_{i=0}^{N} u(x_i)C_i(x) = \Omega(x) + \sum_{i=1}^{N} u(x_i)\Phi_i(x).$$

The interpolation error is bounded for x in $[x_k, x_{k+1}]$ by

$$|W(x) - u(x)| \leq \tfrac{1}{2}|(x - x_k)(x - x_{k+1})| \cdot |u''(\alpha_k)|$$

$$\leq \tfrac{1}{8}(x_{k+1} - x_k)^2 |u''(\alpha_k)|, \qquad x_k \leq \alpha_k \leq x_{k+1}. \quad (5.2.4)$$

To prove this well-known polynomial interpolation error bound, consider the auxiliary function

$$g(s) \equiv W(s) - u(s) - \frac{(s - x_k)(s - x_{k+1})}{(x - x_k)(x - x_{k+1})}[W(x) - u(x)],$$

where x is any (fixed) point between x_k and x_{k+1}. It is easily verified directly that $g(x_k) = g(x_{k+1}) = g(x) = 0$. By the mean value theorem there is a point ξ_1 between x_k and x where g' is zero, and similarly a point ξ_2 between x and x_{k+1} where g' is zero. By applying the mean value theorem to g', it is seen that there must be a point α between ξ_1 and ξ_2 where g'' is zero. That is,

$$g''(\alpha) = W''(\alpha) - u''(\alpha) - \frac{2}{(x - x_k)(x - x_{k+1})}[W(x) - u(x)] = 0.$$

Since W is a linear polynomial in $[x_k, x_{k+1}]$, $W'' \equiv 0$, so

$$u(x) - W(x) = \tfrac{1}{2}(x - x_k)(x - x_{k+1})u''(\alpha),$$

from which 5.2.4 follows.

The maximum interpolation error is therefore bounded by

$$\| W - u \|_\infty \leq \tfrac{1}{8} \Delta x_{\max}^2 \| u'' \|_\infty, \qquad (5.2.5)$$

where $\Delta x_{\max}$ is the maximum of the subinterval widths, $|x_{k+1} - x_k|$, and the infinity norm means the maximum over the interval $(0, 1)$.

Assuming u'' is bounded in $(0, 1)$, there exists a trial function W that approximates u to $O(\Delta x_{\max}^2)$ accuracy, and by the comments in Section 5.1, it is expected that the maximum Galerkin error $\| U - u \|_\infty$ will be $O(\Delta x_{\max}^2)$ also. Of course, this has already been verified by noting the equivalence of this finite element method with a centered finite difference method.

Actually, the interpolation error in the "energy" norm, which involves derivatives of W and u, can be shown to be only $O(\Delta x)$ (cf. Problem 1b), so that, strictly speaking, all 5.1.7 tells us is that the Galerkin error in the energy norm is $O(\Delta x)$. But, in practice, it may usually be assumed that the norm used is unimportant, as confirmed in this case.

Obviously, a lot of effort has been spent in the preceding example deriving formulas that are much easier to derive using finite differences. But it is now clear that formulas of higher-order accuracy can be derived by using the finite element method, if the functions $\Omega, \Phi_1, \ldots, \Phi_M$ are chosen to be higher-order piecewise polynomials.

In addition, formula 5.2.5 shows that the nodes x_i could have been spaced unevenly and the error would still be $O(\Delta x_{\max}^2)$ (Problem 2). The reason why such a nonuniform mesh might be desirable is suggested by 5.2.4. If u'' is much larger in a certain portion of the interval than in other parts, the nodes x_k can be spaced closer together there to offset the increase in u'' (see Section 5.4).

Since the differential equation 5.2.1 involves second derivatives of the solution, it is perhaps surprising that the discontinuity of the first (and second) derivatives of the trial functions does not cause trouble for the Galerkin method. In fact, when a second-order problem is approximated, all that is required of the Galerkin trial functions is continuity. (The first derivative must also be continuous when a fourth-order problem is solved.) An intuitive explanation of this requirement is as follows. The Galerkin method solves the approximation 5.1.4 to the weak formulation 5.1.2 of a second-order partial differential equation, so the Galerkin equations involve only first derivatives of the trial functions. These equations supply information about the first derivatives of U, but this information is useless if U is discontinuous, since only if U is continuous is it possible to obtain information about U from its first derivatives. (Recall that $U(x) = U(a) + \int_a^x U'(s)\,ds$ is valid only when U is continuous.)

5.3   An Example Using Cubic Hermite Trial Functions

Now consider the nonlinear, nonsymmetric problem 4.2.1, repeated here for convenience:

$$u_{xx} - u_x^2 - u^2 + u + 1 = 0, \tag{5.3.1}$$

$$u(0) = 0.5,$$

$$u(\pi) = -0.5.$$

One solution is $u(x) = \sin(x + \pi/6)$. Since this cannot be put into the form 5.1.1, the Galerkin finite element equations must be rederived for this problem.

   Following the procedure established in Section 5.1, the weak formulation (which involves only first derivatives) for this problem is found by multiplying the partial differential equation by an arbitrary $\Phi(x)$ that vanishes on $\partial R_1(x = 0, \pi)$, and by integrating over $[0, \pi]$. (∂R_2 is the empty set now.) This gives

$$0 = \int_0^\pi \{u'' - (u')^2 - u^2 + u + 1\}\Phi \, dx.$$

Integrating by parts, using $\Phi(0) = \Phi(\pi) = 0$,

$$0 = \int_0^\pi \{-u'\Phi' - (u')^2\Phi - u^2\Phi + u\Phi + \Phi\} \, dx. \tag{5.3.2}$$

The Galerkin equations result from the requirement that the approximate solution U satisfy the weak formulation 5.3.2 for $\Phi = \Phi_1, \ldots, \Phi_M$:

$$f_k \equiv \int_0^\pi \{-U'\Phi_k' - (U')^2\Phi_k - U^2\Phi_k + U\Phi_k + \Phi_k\} \, dx = 0. \tag{5.3.3}$$

For a general nonsymmetric problem such as this one, there is no longer any guarantee of convergence, but it is hoped that making the partial differential equation residual "orthogonal" to a larger and larger space of functions will cause the partial differential equation to be more and more nearly satisfied. The (essential) boundary conditions are satisfied automatically by the proper choice of basis functions.

Since $U(x) = \Omega(x) + \sum a_j \Phi_j(x)$, 5.3.3 is a system of M *nonlinear* equations for the M unknowns $a_1, \ldots, a_M$. Newton's method will be used to solve this system, so the elements of the Jacobian of $\mathbf{f}$ must be calculated:

$$\frac{\partial f_k}{\partial a_i} = \int_0^\pi \left[-\Phi_k' \Phi_i' - 2U'\Phi_k \Phi_i' - 2U\Phi_k \Phi_i + \Phi_k \Phi_i \right] dx. \qquad (5.3.4)$$

The functions $\Omega, \Phi_1, \ldots, \Phi_M$ will be chosen from the "cubic Hermite" basis functions, defined by:

$$
\begin{aligned}
H_k(x) &= 3\left[\frac{x - x_{k-1}}{x_k - x_{k-1}}\right]^2 - 2\left[\frac{x - x_{k-1}}{x_k - x_{k-1}}\right]^3, & &\text{for } x_{k-1} \le x \le x_k, \\
&= 3\left[\frac{x_{k+1} - x}{x_{k+1} - x_k}\right]^2 - 2\left[\frac{x_{k+1} - x}{x_{k+1} - x_k}\right]^3, & &\text{for } x_k < x \le x_{k+1}, \qquad (5.3.5a) \\
&= 0, & &\text{elsewhere,} \\[1em]
S_k(x) &= -\frac{(x - x_{k-1})^2}{(x_k - x_{k-1})} + \frac{(x - x_{k-1})^3}{(x_k - x_{k-1})^2}, & &\text{for } x_{k-1} \le x \le x_k, \\
&= \frac{(x_{k+1} - x)^2}{(x_{k+1} - x_k)} - \frac{(x_{k+1} - x)^3}{(x_{k+1} - x_k)^2}, & &\text{for } x_k < x \le x_{k+1}, \qquad (5.3.5b) \\
&= 0, & &\text{elsewhere,}
\end{aligned}
$$

where $x_{-1} < 0 = x_0 < x_1 < \cdots < x_N = \pi < x_{N+1}$.

These functions are piecewise cubic polynomials and are constructed so that they and their first derivatives are continuous. It is also easy to verify directly that

$$
\begin{aligned}
H_k(x_i) &= \delta_{ki}, & H_k'(x_i) &= 0, \\
S_k(x_i) &= 0, & S_k'(x_i) &= \delta_{ki},
\end{aligned}
\qquad (5.3.6)
$$

where δ_{ki} is zero unless $i = k$, in which case it is 1. These functions are drawn in Figure 5.3.1.

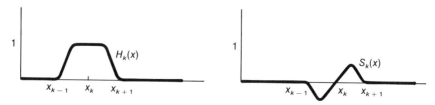

Figure 5.3.1

We choose as our trial functions the following:

$$\Omega(x) = u(x_0)H_0(x) + u(x_N)H_N(x) = \tfrac{1}{2}H_0(x) - \tfrac{1}{2}H_N(x)$$

and

$$\{\Phi_1, \ldots, \Phi_{2N}\} = \{S_0, H_1, S_1, \ldots, H_{N-1}, S_{N-1}, S_N\},$$

so the number of unknowns and equations is $M = 2N$.

It is easy to verify by using 5.3.6 that Ω satisfies the boundary conditions at $x = 0$ and $x = \pi$ and that each Φ_k vanishes at both endpoints, so that these are acceptable trial functions.

Before actually solving 5.3.3, however, let us see how closely we can approximate $u(x)$ with functions in the trial space, which will give us an estimate of the Galerkin method error, according to the comments in Section 5.1. The interpolation function $W(x)$ defined by

$$W(x) = \sum_{k=0}^{N} u(x_k)H_k(x) + \sum_{k=0}^{N} u'(x_k)S_k(x)$$

$$= \Omega(x) + \sum_{k=1}^{N-1} u(x_k)H_k(x) + \sum_{k=0}^{N} u'(x_k)S_k(x)$$

interpolates to the solution $u(x)$ and to its first derivative at $x_0, \ldots, x_N$, as is easily verified by using 5.3.6, and it is a member of the trial space. It will be shown that the difference between W and the true solution is bounded for x in $[x_k, x_{k+1}]$ by

$$|W(x) - u(x)| \le \tfrac{1}{24}(x - x_k)^2(x - x_{k+1})^2|u^{iv}(\alpha_k)|$$

$$\le \tfrac{1}{384}(x_{k+1} - x_k)^4|u^{iv}(\alpha_k)|, \qquad x_k \le \alpha_k \le x_{k-1}. \tag{5.3.7}$$

Thus if u is smooth, there is a member W of the approximating space such that

$$\|W - u\|_\infty \le \tfrac{1}{384}\Delta x_{max}^4 \|u^{iv}\|_\infty.$$

Therefore, it is reasonable to expect the Galerkin error to also be $O(\Delta x_{max}^4)$.

To prove the interpolation error bound 5.3.7, consider the auxiliary function

$$g(s) \equiv W(s) - u(s) - \frac{(s - x_k)^2(s - x_{k+1})^2}{(x - x_k)^2(x - x_{k+1})^2}[W(x) - u(x)],$$

where x is a fixed point in (x_k, x_{k+1}). Its derivative is

$$g'(s) = W'(s) - u'(s) - \frac{2(s - x_k)(s - x_{k+1})(2s - x_k - x_{k+1})}{(x - x_k)^2(x - x_{k+1})^2} [W(x) - u(x)].$$

It is easily verified directly that the function g is zero at $s = x_k$, x, and x_{k+1}, and that g' is zero at $s = x_k$ and x_{k+1}. (Recall that $W = u$ and $W' = u'$ at the nodes.) By the mean value theorem, g' is also zero at some point ξ_1 between x_k and x, and at some point ξ_2 between x and x_{k+1}. Applying the mean value theorem repeatedly,

$g = 0$	at	x_k	x	x_{k+1}
$g' = 0$	at	x_k	ξ_1 $\quad$ ξ_2	x_{k+1}
$g'' = 0$	at	Γ_1	Γ_2	Γ_3
$g''' = 0$	at	β_1	β_2	
$g^{iv} = 0$	at		α	

So, there is some point α in $[x_k, x_{k+1}]$ such that $g^{iv}(\alpha) = 0$. Thus,

$$g^{iv}(\alpha) = W^{iv}(\alpha) - u^{iv}(\alpha) - \frac{24}{(x - x_k)^2(x - x_{k+1})^2} [W(x) - u(x)] = 0.$$

Since W is a cubic polynomial in the interval, $W^{iv} \equiv 0$ and

$$u(x) - W(x) = \tfrac{1}{24}(x - x_k)^2(x - x_{k+1})^2 u^{iv}(\alpha(x)), \qquad (5.3.8)$$

from which 5.3.7 follows.

Now we are ready to solve the nonlinear system 5.3.3 by using the cubic Hermite basis functions specified above. Newton's method for solving a nonlinear system of equations is defined by the iteration

$$\mathbf{a}^{n+1} = \mathbf{a}^n - \boldsymbol{\Delta}^n,$$

$$J^n \boldsymbol{\Delta}^n = \mathbf{f}^n,$$

where the components of the vector $\mathbf{f}^n$ and its Jacobian matrix J^n are given by formulas 5.3.3 and 5.3.4. The superscript n means that the functions are evaluated at $\mathbf{a}^n = (a_1^n, \ldots, a_M^n)$.

When the Hermite basis functions are used, the Jacobian matrix is banded, with half-bandwidth equal to 3. The nonzero structure of J is shown in Figure 5.3.2, for $N = 8$ ($M = 16$).

$$
\begin{bmatrix}
X & X & X & & & & & & & & & & & \\
X & X & X & X & X & & & & & & & & & \\
X & X & X & X & X & & & & & & & & & \\
& & & X & X & X & X & X & X & & & & & \\
& & & X & X & X & X & X & X & & & & & \\
& & & & & X & X & X & X & X & X & & & \\
& & & & & X & X & X & X & X & X & & & \\
& & & & & & & X & X & X & X & X & X & \\
& & & & & & & X & X & X & X & X & X & \\
& & & & & & & & & X & X & X & X & X & X \\
& & & & & & & & & X & X & X & X & X & X \\
& & & & & & & & & & & X & X & X & X & X \\
& & & & & & & & & & & X & X & X & X & X \\
& & & & & & & & & & & & & X & X & X
\end{bmatrix}
$$

Figure 5.3.2
Nonzero Structure of Jacobian, Using Cubic Hermites

The integrals in 5.3.3 and 5.3.4 must clearly be evaluated numerically. Two different numerical integration schemes of the form

$$
\int_0^\pi g(x)\,dx \approx \sum_{j=0}^{N-1} h_j[w_1 g(x_j + s_1 h_j) + w_2 g(x_j + s_2 h_j) + w_3 g(x_j + s_3 h_j)]
$$

were tried ($h_j \equiv x_{j+1} - x_j$), namely

Simpson's rule: $w_1 = 0.1666666666666667$ $s_1 = 0.0000000000000000$
$w_2 = 0.6666666666666667$ $s_2 = 0.5000000000000000$
$w_3 = 0.1666666666666667$ $s_3 = 1.0000000000000000$

Gauss 3-point rule: $w_1 = 0.2777777777777778$ $s_1 = 0.1127016653792583$
$w_2 = 0.4444444444444444$ $s_2 = 0.5000000000000000$
$w_3 = 0.2777777777777778$ $s_3 = 0.8872983346207417.$

Simpson's rule is known to be exact when $g(x)$ is a polynomial of degree 3, while the Gauss 3-point rule is exact for polynomials of degree 5. The nodes were spaced uniformly, $x_j = j\Delta x$ ($\Delta x = \pi/N$).

When an initial guess of $U(x_k) = \frac{1}{2} - x_k/\pi$ ($U'(x_k) = -1/\pi$), chosen to satisfy the boundary conditions, was used, Newton's method converged in five to ten iterations for each value of Δx used. The maximum error (after convergence) for three values of Δx is shown in Table 5.3.1. "Maximum error" here means the maximum over 100 points uniformly distributed in $[0, \pi]$.

Table 5.3.1

Maximum Errors for Nonlinear Problem

N ($\Delta x = \pi/N$)	Maximum error using	
	Gauss 3-point rule	Simpson's rule
4	0.3900E − 3	0.6487E − 2
8	0.2962E − 4	0.6280E − 3
16	0.1961E − 5	0.6828E − 4

For the Gauss 3-point rule case, when Δx is halved, the error decreases by factors of 13.2 and 15.1, suggesting an error behavior of $O(\Delta x^4)$, as predicted by the theory. When Simpson's rule is used, the error decreases by factors of 10.3 and 9.2, suggesting an error of only about $O(\Delta x^3)$. This is consistent with the theory, because it is known [Strang and Fix, 1973—see also Problem 1] that an integration rule exact for polynomials of degree $2n - 2$ must be used, where n is the degree of the piecewise polynomial space used ($n = 3$ in the case of cubic Hermites), if the overall order of convergence of the Galerkin method is not to be diminished. Thus an integration rule exact for polynomials of degree 4 is required, and Simpson's rule is not good enough. Recall that in the example of Section 5.2, polynomials of degree $n = 1$ were used so that a rule exact for constants is all that is required. The trapezoid rule is exact for linear polynomials, which explains why the order of convergence was not ruined by the numerical integration in that problem.

The FORTRAN77 program used to solve 5.3.1 is shown in Figure 5.3.3. Those subprograms that are general enough to be useful when solving other problems by using cubic Hermite basis functions are isolated in Figure 5.3.3b. The band solver LBAND (Figure 0.4.3) is called in the main program, to solve the linear system each Newton iteration.

Some efficiency has been sacrificed in the interest of readability. In particular, subprogram USOL could be improved by limiting the range of I in DO loop 10 to $I = \text{MAX}(K - 3, 1)$ to $\text{MIN}(K + 3, M)$ when USOL is called from function GAUSS, since only for these values of I can $\Phi_I(X)$ be nonzero. A more significant improvement in speed can be obtained by employing an "element-by-element" assembly technique, as discussed in Section 5.6.

The large amount of programming effort involved in solving this problem by using cubic Hermites is surely not justified if a single problem is to be

```
      IMPLICIT DOUBLE PRECISION(A-H,O-Z)
      EXTERNAL FUNJ,FUNF
      CHARACTER*1 HORS
C                                    NSUBS = NUMBER OF SUBINTERVALS
      PARAMETER (NSUBS=4)
C                                    BREAK POINTS ARE XPTS(K), K=0,N
      COMMON /BLKX/ N,XPTS(0:NSUBS)
C                                    UNKNOWNS ARE A(I), I=1,M
      COMMON /BLKA/ M,A(2*NSUBS)
      DIMENSION AJACOB(2*NSUBS,7),F(2*NSUBS)
      DATA PI/3.14159265358979312D0/
      N = NSUBS
      M = 2*NSUBS
      DO 5 K=0,N
         XPTS(K) = K*PI/N
    5 CONTINUE
C                                    INITIAL GUESS FOR NEWTON METHOD
      DO 10 I=1,M
         CALL LOCATE(I,HORS,K)
         IF (HORS.EQ.'H') A(I) = 0.5-XPTS(K)/PI
         IF (HORS.EQ.'S') A(I) = -1.0/PI
   10 CONTINUE
C                                    DO ITMAX NEWTON ITERATIONS
      ITMAX = 15
      DO 40 ITER=1,ITMAX
C                                    CALCULATE F VECTOR AND JACOBIAN
         DO 20 K=1,M
            F(K) = GAUSS(FUNF,K,0)
            DO 15 ISUB=1,7
               I=K-4+ISUB
               IF (I.GE.1.AND.I.LE.M) AJACOB(K,ISUB) = GAUSS(FUNJ,K,I)
   15       CONTINUE
   20    CONTINUE
C                                    SOLVE LINEAR SYSTEM USING BAND SOLVER
         L = 3
         CALL LBAND(AJACOB,F,F,M,M,L)
C                                    UPDATE SOLUTION VECTOR
         DO 25 I=1,M
            A(I) = A(I)-F(I)
   25    CONTINUE
C                                    CALCULATE MAXIMUM ERROR
         ERMAX = 0.0
         DO 30 J=0,100
            X = J*PI/100.D0
            ERR = ABS(USOL(0,X) - SIN(X+PI/6.D0))
            ERMAX = MAX(ERMAX,ERR)
   30    CONTINUE
         PRINT 35,ITER,ERMAX
   35    FORMAT (I5,E15.5)
```

(continued)

```
   40 CONTINUE
      STOP
      END
      FUNCTION FUNF(X,U,UX,UXX,PK,PKX,PKXX)
      IMPLICIT DOUBLE PRECISION(A-H,O-Z)
C                          INTEGRAND FOR K-TH ELEMENT OF F VECTOR
      FUNF = -UX*PKX - UX**2*PK - U**2*PK + U*PK + PK
      RETURN
      END
      FUNCTION FUNJ(X,U,UX,UXX,PK,PKX,PKXX,PI,PIX,PIXX)
      IMPLICIT DOUBLE PRECISION(A-H,O-Z)
C                          INTEGRAND FOR (K,I)-TH ELEMENT OF
C                          JACOBIAN MATRIX
      FUNJ = -PKX*PIX - 2*UX*PK*PIX - 2*U*PK*PI + PK*PI
      RETURN
      END
      SUBROUTINE LOCATE(I,HORS,K)
      IMPLICIT DOUBLE PRECISION(A-H,O-Z)
      CHARACTER*1 HORS
      COMMON /BLKA/ M,A(1)
C                          DETERMINE THE TYPE ('H' OR 'S') OF BASIS
C                          FUNCTION I, AND THE POINT (XPTS(K)) ON WHICH
C                          IT IS CENTERED.
      IF (MOD(I,2).EQ.0) HORS = 'H'
      IF (MOD(I,2).EQ.1) HORS = 'S'
      IF (I.EQ.M) HORS = 'S'
      K = I/2
      RETURN
      END
      FUNCTION OMEGA(IDER,X)
      IMPLICIT DOUBLE PRECISION(A-H,O-Z)
      COMMON /BLKX/ N,XPTS(0:1)
C                          EVALUATE OMEGA (IDER=0) OR ITS IDER-TH
C                          DERIVATIVE AT X
      OMEGA = 0.5*HERM('H',0,IDER,X) - 0.5*HERM('H',N,IDER,X)
      RETURN
      END
```

Figure 5.3.3a
FORTRAN77 Program to Solve Nonlinear Problem

solved, since equivalent precision can be obtained by using finite differences (as in Section 4.2) on a much finer grid, and the increased computer cost is rarely important for one-dimensional problems. On the other hand, if one is writing a subroutine that will be used repeatedly, the extra effort may be worth it. If a nonuniform grid is required, the finite element approach becomes even more attractive.

```
      FUNCTION USOL(IDER,X)
      IMPLICIT DOUBLE PRECISION(A-H,O-Z)
      COMMON /BLKA/ M,A(1)
C                              EVALUATE THE APPROXIMATE SOLUTION (IDER=0)
C                              OR ITS IDER-TH DERIVATIVE AT X
      USOL = OMEGA(IDER,X)
      DO 10 I=1,M
         USOL = USOL + A(I)*PHI(I,IDER,X)
   10 CONTINUE
      RETURN
      END
      FUNCTION PHI(I,IDER,X)
      IMPLICIT DOUBLE PRECISION(A-H,O-Z)
      CHARACTER*1 HORS
C                              EVALUATE THE I-TH BASIS FUNCTION (IDER=0)
C                              OR ITS IDER-TH DERIVATIVE AT X
      CALL LOCATE(I,HORS,K)
      PHI = HERM(HORS,K,IDER,X)
      RETURN
      END
      FUNCTION HERM(HORS,K,IDER,X)
      IMPLICIT DOUBLE PRECISION(A-H,O-Z)
      CHARACTER*1 HORS
      COMMON /BLKX/ N,XPTS(0:1)
C                              EVALUATE AT X THE IDER-TH DERIVATIVE OF THE
C                              CUBIC HERMITE 'H' OR 'S' FUNCTION CENTERED AT
C                              XPTS(K).
      XK = XPTS(K)
      HH = XPTS(N)-XPTS(0)
      IF (K.GT.0) XKM1 = XPTS(K-1)
      IF (K.LE.0) XKM1 = XPTS(0)-HH
      IF (K.LT.N) XKP1 = XPTS(K+1)
      IF (K.GE.N) XKP1 = XPTS(N)+HH
      HERM = 0.0
      IF (X.LT.XKM1) RETURN
      IF (X.GT.XKP1) RETURN
      IF (X.LE.XK) THEN
         H = XK-XKM1
         S = (X-XKM1)/H
         IF (HORS.EQ.'H') THEN
            IF (IDER.EQ.0) HERM = 3.0*S**2 - 2.0*S**3
            IF (IDER.EQ.1) HERM = 6.0*S/H - 6.0*S**2/H
            IF (IDER.EQ.2) HERM = 6.0/H**2 - 12.0*S/H**2
         ELSE
            IF (IDER.EQ.0) HERM = -H*S**2 + H*S**3
            IF (IDER.EQ.1) HERM = -2.0*S + 3.0*S**2
            IF (IDER.EQ.2) HERM = -2.0/H + 6.0*S/H
         ENDIF
```

(continued)

```fortran
      ELSE
        H = XKP1-XK
        S = (XKP1-X)/H
        IF (HORS.EQ.'H') THEN
          IF (IDER.EQ.0) HERM = 3.0*S**2 - 2.0*S**3
          IF (IDER.EQ.1) HERM = -6.0*S/H + 6.0*S**2/H
          IF (IDER.EQ.2) HERM = 6.0/H**2 - 12.0*S/H**2
        ELSE
          IF (IDER.EQ.0) HERM = H*S**2 - H*S**3
          IF (IDER.EQ.1) HERM = -2.0*S + 3.0*S**2
          IF (IDER.EQ.2) HERM = 2.0/H - 6.0*S/H
        ENDIF
      ENDIF
      RETURN
      END
      FUNCTION GAUSS(F,K,I)
      IMPLICIT DOUBLE PRECISION(A-H,O-Z)
      CHARACTER*1 HORS
      DIMENSION S(3),W(3)
      COMMON /BLKX/ N,XPTS(0:1)
C                          USE THE GAUSS 3 POINT FORMULA TO
C                          INTEGRATE F(X)
      DATA S
     & /0.1127016653792583D0,0.5000000000000000D0,0.8872983346207417D0/
      DATA W
     & /0.2777777777777778D0,0.4444444444444444D0,0.2777777777777778D0/
      GAUSS = 0.0
C                          THE FUNCTION F INVOLVES PHI(K,...), WHICH
C                          IS NONZERO ONLY FROM XPTS(IK-1) TO XPTS(IK+1)
      CALL LOCATE(K,HORS,IK)
      DO 10 J = MAX(IK,1),MIN(IK+1,N)
        H = XPTS(J)-XPTS(J-1)
        DO 5 JX=1,3
          X = XPTS(J-1)+S(JX)*H
          IF (I.EQ.0) THEN
            GAUSS = GAUSS + H*W(JX)*F(X,USOL(0,X),USOL(1,X),USOL(2,X),
     &                      PHI(K,0,X),PHI(K,1,X),PHI(K,2,X))
          ELSE
            GAUSS = GAUSS + H*W(JX)*F(X,USOL(0,X),USOL(1,X),USOL(2,X),
     &                      PHI(K,0,X),PHI(K,1,X),PHI(K,2,X),
     &                      PHI(I,0,X),PHI(I,1,X),PHI(I,2,X))
          ENDIF
    5   CONTINUE
   10 CONTINUE
      RETURN
      END
```

Figure 5.3.3b
Problem Independent Portion of FORTRAN77 Program

5.4 A Singular Example

Let us look again at the singular problem 4.3.1, which can be put into the form 5.1.1 by dividing by x^2:

$$u_{xx} - 0.11u/x^2 = 0,$$
$$u(0) = 0, \tag{5.4.1}$$
$$u_x(1) = 1.1.$$

Since the solution $u(x) = x^{1.1}$ has unbounded second derivatives near $x = 0$, the finite difference solution calculated earlier with a uniform grid converged slowly (see Table 4.3.1). We will therefore use cubic Hermite basis functions with an appropriately chosen *nonuniform* mesh.

We choose $\Omega \equiv 0$ and $\{\Phi_1, \ldots, \Phi_{2N+1}\} = \{S_0, H_1, S_1, \ldots, H_N, S_N\}$, where H_i, S_i are the cubic Hermite functions defined in 5.3.5. All of the Φ_i are zero on $\partial R_1(x = 0)$ as required, since H_0 is not in the basis, and Ω certainly satisfies the essential boundary condition there. Alternatively, we could choose $\Omega = 1.1S_N$, $\{\Phi_1, \ldots, \Phi_{2N}\} = \{S_0, H_1, S_1, \ldots, H_{N-1}, S_{N-1}, H_N\}$, and then the natural boundary condition is imposed exactly. The first approach generalizes more easily to other trial spaces, however, where it may not be possible to impose the natural boundary condition directly.

For problem 5.4.1, $D = 1$, $a = 0.11/x^2$, $f = 0$, $r = 0$, $p = 0$, $q = 1.1$, $\partial R_1 = \{x = 0\}$, and $\partial R_2 = \{x = 1\}$ (see 5.1.1). Formulas 5.1.5 can be used to calculate the elements of the matrix and right-hand vector in the finite element linear system, provided it is remembered that R is now the interval $[0, 1]$ and ∂R_2 is the single point $x = 1$. Thus we have

$$A_{ki} = \int_0^1 \left[\Phi_k' \Phi_i' + \frac{0.11}{x^2} \Phi_k \Phi_i \right] dx,$$
$$b_k = 1.1\, \Phi_k(1).$$

From the comments in Section 5.1, it seems reasonable to distribute the Hermite break points x_k with the idea of minimizing the difference between the exact solution and its cubic Hermite interpolant (which is a member of the approximating set since it can be expressed in the form $W = \Omega + \Sigma\, a_j \Phi_j$). According to 5.3.7, this interpolation error is

$$\|W - u\|_\infty \leq \frac{1}{384} \max_k \left[(x_{k+1} - x_k)^4 |u^{iv}(\alpha_k)| \right]$$

$$= \frac{1}{384} \max_k \left[|u^{iv}(\alpha_k)|^{1/4}(x_{k+1} - x_k) \right]^4 \qquad x_k \leq \alpha_k \leq x_{k+1}.$$

If we assume that $u^{iv}(x)$ is smooth and that the mesh is fine enough so that u^{iv} is nearly constant in each subinterval, then (approximately)

$$\| W - u \|_\infty \leq \frac{1}{384} \max_k \left[\int_{x_k}^{x_{k+1}} |u^{iv}(x)|^{1/4} \, dx \right]^4.$$

Since the sum of the little integrals is fixed, namely $\int_0^1 |u^{iv}|^{1/4} \, dx$, the maximum of the small integrals is minimized when they are all equal, that is, when

$$\int_0^{x_k} |u^{iv}(x)|^{1/4} \, dx = \frac{k}{N} \int_0^1 |u^{iv}(x)|^{1/4} \, dx, \qquad (5.4.2)$$

in which case,

$$\| W - u \|_\infty \leq \frac{1}{384} \frac{1}{N^4} \left[\int_0^1 |u^{iv}(x)|^{1/4} \, dx \right]^4. \qquad (5.4.3)$$

For this problem, $u(x) = x^{1.1}$, $u^{iv}(x) = 0.1881x^{-2.9}$, and therefore $u^{iv}(x)$ is not bounded, contrary to an assumption used to derive 5.4.3. Nevertheless, the integral in 5.4.3 is finite, suggesting that perhaps if this integral is equidistributed, the "optimal" $O(1/N^4)$ convergence rate (expected for well-behaved solutions using a uniform mesh) may be attained anyway. Therefore, we choose the x_k so that 5.4.2 is satisfied:

$$\int_0^{x_k} (x^{-2.9})^{1/4} \, dx = \frac{k}{N} \int_0^1 (x^{-2.9})^{1/4} \, dx$$

or $x_k = (k/N)^{3.636}$.

The maximum errors resulting when this mesh spacing was used, as well as two others of similar form, are shown in Table 5.4.1. The "convergence rate" means the power of $1/N$ to which the error is experimentally proportional— this is calculated by comparing the errors for $N = 16$ and $N = 32$.

Table 5.4.1
Maximum Errors Using Galerkin Method

N	$x_k = (k/N)^\alpha$		
	$\alpha = 1$ (uniform)	$\alpha = 3.636$	$\alpha = 5$
4	0.8194E − 3	0.1361E − 3	0.3178E − 3
8	0.3754E − 3	0.1034E − 4	0.2508E − 4
16	0.1725E − 3	0.6533E − 6	0.1902E − 5
32	0.6509E − 4	0.4157E − 7	0.1318E − 6
Convergence rate	1.41	3.97	3.85

With a uniform grid, convergence was not much faster than with the uniform grid finite difference formula (Table 4.3.1). However, with $\alpha \geq 3.636$, the optimal $O(1/N^4)$ convergence rate was attained, and $\alpha = 3.636$ gave smaller errors than $\alpha = 5$.

For most problems, of course, we shall not have the advantage of knowing the exact solution, so distributing the grid points optimally may not be possible. However, as the results for $\alpha = 5$ illustrate, if the solution is poorly behaved, any reasonable heuristic grading of the mesh will do better than a uniform one. Ideally, of course, a sophisticated boundary value problem solver should determine the mesh grading adaptively.

Many theoretical and experimental results suggest that optimal-order convergence (in the infinity norm) is possible, with an appropriately graded grid, provided only that the solution is *continuous*!

The FORTRAN77 program used to solve 5.4.1 is shown in Figure 5.4.1. The subprograms of Figure 5.3.3b and the band solver LBAND (Figure 0.4.3) are called by this program and thus must be loaded along with these subprograms.

As mentioned in Section 5.1, the Galerkin method is not the only form of the finite element method in use. Another form is the *collocation* method, in which the solution is expanded as a linear combination of piecewise polynomial basis functions similar to the expansion 5.1.3 used by the Galerkin method. However, rather than requiring that the partial differential equation be satisfied when it is multiplied by Φ_k and integrated, the collocation methods determine the coefficients in the solution expansion by requiring that the partial differential equation and natural boundary condition be satisfied exactly at certain specified points.

To illustrate the use of the collocation method, let us solve again the boundary value problem 5.4.1. In the expansion for the solution, we choose $\Omega = 1.1 S_N$ and $\{\Phi_1, \ldots, \Phi_{2N}\} = \{S_0, H_1, S_1, \ldots, H_{N-1}, S_{N-1}, H_N\}$. Then the approximate solution $U(x) \equiv \Omega(x) + \Sigma \, a_i \Phi_i(x)$ automatically satisfies both the essential and natural boundary conditions, no matter what values are given to the coefficients a_i.

To determine the $2N$ coefficients a_i, we need $2N$ collocation points, or two in each interval: $x_l + \beta_j(x_{l+1} - x_l)$, for $l = 0, \ldots, N - 1, j = 1, 2$. The requirement that $U(x)$ satisfy 5.4.1 at the collocation point z_k can be written $(M = 2N)$

$$\sum_{i=1}^{M} a_i \left[\Phi_i''(z_k) - \frac{0.11 \Phi_i(z_k)}{z_k^2} \right] = -\Omega''(z_k) + \frac{0.11 \Omega(z_k)}{z_k^2}. \qquad (5.4.4)$$

```
      IMPLICIT DOUBLE PRECISION(A-H,O-Z)
      EXTERNAL FUNA
C                                 NSUBS = NUMBER OF SUBINTERVALS
      PARAMETER (NSUBS=4)
C                                 BREAK POINTS ARE XPTS(K), K=0,N
      COMMON /BLKX/ N,XPTS(0:NSUBS)
C                                 UNKNOWNS ARE A(I), I=1,M
      COMMON /BLKA/ M,A(2*NSUBS+1)
      DIMENSION AMAT(2*NSUBS+1,7),B(2*NSUBS+1)
      N = NSUBS
      M = 2*NSUBS+1
      ALPHA = 3.636
      DO 5 K=0,N
         XPTS(K) = (DBLE(K)/N)**ALPHA
    5 CONTINUE
C                                 CALCULATE RIGHT HAND SIDE VECTOR AND
C                                 COEFFICIENT MATRIX
      DO 15 K=1,M
         B(K) = 1.1D0*PHI(K,0,1.D0)
         DO 10 ISUB=1,7
            I = K-4+ISUB
            IF (I.GE.1.AND.I.LE.M) AMAT(K,ISUB) = GAUSS(FUNA,K,I)
   10    CONTINUE
   15 CONTINUE
C                                 SOLVE LINEAR SYSTEM USING BAND SOLVER
      L = 3
      CALL LBAND(AMAT,A,B,M,M,L)
C                                 CALCULATE MAXIMUM ERROR
      ERMAX = 0.0
      DO 20 J=0,100
         X = J/100.D0
         ERR = ABS(USOL(0,X) - X**1.1D0)
         ERMAX = MAX(ERMAX,ERR)
   20 CONTINUE
      PRINT 25, ERMAX
   25 FORMAT (E15.5)
      STOP
      END
      FUNCTION FUNA(X,U,UX,UXX,PK,PKX,PKXX,PI,PIX,PIXX)
      IMPLICIT DOUBLE PRECISION(A-H,O-Z)
C                                 INTEGRAND FOR (K,I)-TH ELEMENT OF
C                                 COEFFICIENT MATRIX
      FUNA = PKX*PIX + 0.11D0/X**2*PK*PI
      RETURN
      END
      SUBROUTINE LOCATE(I,HORS,K)
      IMPLICIT DOUBLE PRECISION(A-H,O-Z)
      CHARACTER*1 HORS
C                                 DETERMINE THE TYPE ('H' OR 'S') OF BASIS
C                                 FUNCTION I, AND THE POINT (XPTS(K)) ON WHICH
C                                 IT IS CENTERED.
```

```
          IF (MOD(I,2).EQ.0) HORS = 'H'
          IF (MOD(I,2).EQ.1) HORS = 'S'
          K = I/2
          RETURN
          END
          FUNCTION OMEGA(IDER,X)
          IMPLICIT DOUBLE PRECISION(A-H,O-Z)
C                          EVALUATE OMEGA (IDER=0) OR ITS IDER-TH
C                          DERIVATIVE AT X
          OMEGA = 0.0
          RETURN
          END
```

Figure 5.4.1

FORTRAN77 Program to Solve Singular Problem, Using Galerkin Method

If the basis functions are ordered as suggested above and the collocation points are taken in ascending order, the resulting linear system 5.4.4 is banded, with half-bandwidth $L = 2$. While the coefficient matrix for the linear system generated earlier (when the Galerkin finite element method was applied to this problem) was positive-definite, the coefficient matrix for the collocation linear system is not even symmetric, as it normally is not. Nevertheless, the band solver of Figure 0.4.3, which does no pivoting, was used successfully on this problem. With $\beta_1 = 0.25$, $\beta_2 = 0.75$, the results were as shown in Table 5.4.2.

Although we have not derived any error bounds for the collocation method, it seems reasonable to expect that, as for the Galerkin method, the error is closely related to the distance from the trial space to the exact solution. That being the case, we should expect the error to be of the same order, $O(1/N^4)$, as obtained by the Galerkin method using the same trial

Table 5.4.2

Maximum Errors, Using Collocation Method

	$x_k = (k/N)^\alpha$		
N	$\alpha = 1$ (uniform)	$\alpha = 3.636$	$\alpha = 5$
4	$0.1694E - 2$	$0.1883E - 2$	$0.2759E - 2$
8	$0.7833E - 3$	$0.5348E - 3$	$0.8168E - 3$
16	$0.3633E - 3$	$0.1392E - 3$	$0.2139E - 3$
32	$0.1683E - 3$	$0.3522E - 4$	$0.5414E - 4$
Convergence rate	1.11	1.98	1.98

functions and the same break points, but the results show a convergence rate of only $O(1/N^2)$ when $\alpha = 3.636$ and $\alpha = 5$. The problem is not the singularity, but the choice of collocation points. To recover the optimal $O(1/N^4)$ rate, it has been shown that the collocation points must be the sample points for the Gauss 2-point quadrature rule, i.e., $\beta_1 = 0.5 - 0.5/\sqrt{3}$, $\beta_2 = 0.5 + 0.5/\sqrt{3}$. This is confirmed by Table 5.4.3, which shows the results when these collocation points were used.

Table 5.4.3
Maximum Errors, Using Collocation Method
With Optimal Collocation Points

N	$\alpha = 1$ (uniform)	$x_k = (k/N)^\alpha$ $\alpha = 3.636$	$\alpha = 5$
4	$0.9191E - 3$	$0.2613E - 3$	$0.5905E - 3$
8	$0.4288E - 3$	$0.2267E - 4$	$0.5939E - 4$
16	$0.1613E - 3$	$0.1676E - 5$	$0.4566E - 5$
32	$0.7504E - 4$	$0.1145E - 6$	$0.3144E - 6$
Convergence rate	1.10	3.87	3.86

The FORTRAN77 collocation program used is shown in Figure 5.4.2. The subprograms of Figure 5.3.3b (GAUSS may be omitted) and the band solver LBAND (Figure 0.4.3) are called by this program, and thus must be loaded along with these subprograms.

Clearly, the piecewise linear functions of Section 5.2, which have discontinuous first derivatives, could not be used to solve a second-order problem such as 5.4.1 by using a collocation method, as the second derivatives of all the basis functions are identically zero. In fact, the trial functions used by a collocation method to solve an nth order problem must have continuous $(n - 1)$st derivatives. An intuitive explanation for this is that the collocation equations supply information about the nth derivatives of U, but this information is useless unless U and its first $n - 1$ derivatives are continuous, since only if $U^{(n-1)}$ is continuous is it possible to obtain information about $U^{(n-1)}$ from $U^{(n)}$, and so on, back to U itself (see Section 5.2). Thus for the second-order problem 5.4.1, trial functions with continuous first derivatives are required, a requirement that is satisfied by our cubic Hermite trial functions. When irregular multi-dimensional elements such as triangles or tetrahedra are used, it is difficult (but not impossible) to construct piecewise

```
      IMPLICIT DOUBLE PRECISION(A-H,O-Z)
C                             NSUBS = NUMBER OF SUBINTERVALS
      PARAMETER (NSUBS=4)
C                             BREAK POINTS ARE XPTS(K), K=0,N
      COMMON /BLKX/ N,XPTS(0:NSUBS)
C                             UNKNOWNS ARE A(I), I=1,M
      COMMON /BLKA/ M,A(2*NSUBS)
      DIMENSION AMAT(2*NSUBS,5),B(2*NSUBS),BETA(2)
      N = NSUBS
      M = 2*NSUBS
      ALPHA = 3.636
      DO 5 K=0,N
        XPTS(K) = (DBLE(K)/N)**ALPHA
    5 CONTINUE
C                             CALCULATE RIGHT HAND SIDE VECTOR AND
C                             COEFFICIENT MATRIX
      BETA(1) = 0.5 - 0.5/SQRT(3.D0)
      BETA(2) = 0.5 + 0.5/SQRT(3.D0)
      DO 15 L=0,N-1
      DO 15 J=1,2
        K = 2*L+J
        ZK = XPTS(L) + BETA(J)*(XPTS(L+1)-XPTS(L))
        B(K) = -OMEGA(2,ZK) + 0.11D0*OMEGA(0,ZK)/ZK**2
        DO 10 ISUB=1,5
          I = K-3+ISUB
          IF (I.GE.1.AND.I.LE.M) AMAT(K,ISUB) =
     &                 PHI(I,2,ZK) - 0.11D0*PHI(I,0,ZK)/ZK**2
   10   CONTINUE
   15 CONTINUE
C                             SOLVE LINEAR SYSTEM USING BAND SOLVER
      L = 2
      CALL LBAND(AMAT,A,B,M,M,L)
C                             CALCULATE MAXIMUM ERROR
      ERMAX = 0.0
      DO 20 J=0,100
        X = J/100.D0
        ERR = ABS(USOL(0,X) - X**1.1D0)
        ERMAX = MAX(ERMAX,ERR)
   20 CONTINUE
      PRINT 25, ERMAX
   25 FORMAT (E15.5)
      STOP
      END
      SUBROUTINE LOCATE(I,HORS,K)
      IMPLICIT DOUBLE PRECISION(A-H,O-Z)
      CHARACTER*1 HORS
C                             DETERMINE THE TYPE ('H' OR 'S') OF BASIS
C                             FUNCTION I, AND THE POINT (XPTS(K)) ON WHICH
```

(continued)

216 THE FINITE ELEMENT METHOD

```
C                        IT IS CENTERED.
      IF (MOD(I,2).EQ.0) HORS = 'H'
      IF (MOD(I,2).EQ.1) HORS = 'S'
      K = I/2
      RETURN
      END
      FUNCTION OMEGA(IDER,X)
      IMPLICIT DOUBLE PRECISION(A-H,O-Z)
      COMMON /BLKX/ N,XPTS(0:1)
C                        EVALUATE OMEGA (IDER=0) OR ITS IDER-TH
C                        DERIVATIVE AT X
      OMEGA = 1.1D0*HERM('S',N,IDER,X)
      RETURN
      END
```

Figure 5.4.2
FORTRAN77 Program to Solve Singular Problem, Using Collocation Method

polynomial basis functions with continuous first derivatives. On a triangulation, polynomials of degree at least five must be used if continuity of the first derivatives is imposed [Strang and Fix, 1973, Section 1.9].

Although collocation methods reduce symmetric differential equations to nonsymmetric linear systems (which is particularly unfortunate when an eigenvalue problem is solved), and although they require more continuity of the basis functions, there is no numerical integration to perform, so the programming effort is reduced compared to Galerkin methods.

5.5   Linear Triangular Elements

The finite element method is useful, although rarely indispensable, for one-dimensional problems. But it is almost a necessity for multi-dimensional problems in odd-shaped regions.

In one-dimensional problems, the "finite elements," on which the trial functions reduce to polynomials, are *intervals* $[x_k, x_{k+1}]$. In two dimensions, the finite elements are usually triangles or quadrilaterals. We shall consider here only the simplest two-dimensional element, the triangle, and its use in solving the two-dimensional version of the partial differential equation 5.1.1.

Any polygonal region can be "triangulated"—divided into triangles—and even a curved two-dimensional region can be approximately triangulated (see Figure 5.5.1). For reasons that will be clear later, it is required that the triangle edges in such a triangulation "match up," that is, that a vertex of one

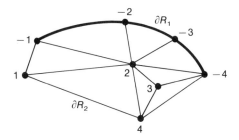

Figure 5.5.1

triangle cannot touch another one in a point that is not a vertex of the second triangle.

A set of piecewise linear functions can be defined on this triangulation as follows. First, the vertices (x_i, y_i) are numbered, using negative integers for those that lie on ∂R_1 (the curved boundary in Figure 5.5.1) and positive integers for the others; then $S_i(x, y)$ is defined to be equal to 1 at vertex number i and 0 at the other vertices. In the interior of each triangle, $S_i(x, y)$ is equal to the linear polynomial (of the form $a + bx + cy$) that takes these prescribed values at the vertices of the triangle.

The functions in the Galerkin expansion 5.1.3 for the approximate solution of 5.1.1 are defined by

$$\Omega(x, y) = \sum_{i<0} r(x_i, y_i) S_i(x, y),$$

$$\Phi_i(x, y) = S_i(x, y), \qquad \text{for } i > 0.$$

It is easy to see that $\Phi_i = 0$ and $\Omega = r$ at every *vertex* on ∂R_1. If ∂R_1 is made up of straight line segments, then clearly, $\Phi_i = 0$ at every point on this boundary, while if it is curved, $\Phi_i = 0$ is satisfied only in the limit as the triangulation is refined. Likewise, $\Omega = r$ holds only in the limit, unless ∂R_1 is polygonal and r happens to be linear.

Note that $S_i(x, y)$ is continuous across a boundary between triangles, because the two neighboring linear functions reduce to straight lines on this boundary, and since both lines have the same values at the extremes (the vertices) of this edge, they are the same line. Although the derivatives of S_i are not continuous across triangle boundaries, continuity is all that is required of the Galerkin method trial functions, as discussed in Section 5.2.

Let us see how closely the solution $u(x, y)$ can be approximated by functions in the trial space of piecewise linear functions; this will give us an estimate of the Galerkin method error, according to 5.1.7. The function $W(x, y)$ that interpolates to u at the vertices, defined by

$$W(x, y) = \sum_{\text{all } i} u(x_i, y_i)S_i(x, y) = \Omega(x, y) + \sum_{i > 0} u(x_i, y_i)\Phi_i(x, y),$$

is a member of the trial set, and we shall show that it is an $O(h_{\text{max}}^2)$ approximation to u, where h_{max} is the longest edge of any triangle in the triangulation.

To show this, first define $u_T(x, y)$ to be, in each triangle, the linear Taylor polynomial interpolant to $u(x, y)$ at the midpoint (x_0, y_0) of that triangle. Then for any (x, y) in this triangle,

$$u(x, y) = u_T(x, y) + \tfrac{1}{2}u_{xx}(\alpha, \beta)(x - x_0)^2$$
$$+ \tfrac{1}{2}u_{yy}(\alpha, \beta)(y - y_0)^2 + u_{xy}(\alpha, \beta)(x - x_0)(y - y_0),$$

where (α, β) is some point on the line between (x_0, y_0) and (x, y). Then

$$\|u_T - u\|_\infty \leq \max_k \; [2h_k^2(D_k^2 u)] \equiv Z,$$

where the infinity norm means the maximum over the entire triangulation, $D_k^2 u$ is a bound on the absolute values of the three second derivatives of u in triangle T_k, and h_k is the diameter (longest side) of T_k.

u_T differs from u by at most Z at any point in triangle T_k including, naturally, its vertices. Since $W = u$ at the vertices, $|u_T - W| \leq Z$ at the vertices. But $u_T(x, y) - W(x, y)$ is a linear function in T_k, and a linear function must take its largest and smallest values in T_k at the vertices. So $|u_T - W| \leq Z$ throughout the (arbitrary) triangle T_k and thus throughout the entire region. So, finally,

$$\|W - u\|_\infty \leq \|W - u_T\|_\infty + \|u_T - u\|_\infty \leq Z + Z$$

or

$$\|W - u\|_\infty \leq 4 \max_k \; [h_k^2(D_k^2 u)]. \tag{5.5.1}$$

If $h_{\text{max}} \equiv \max h_k$, over all triangles T_k, and the second derivatives of the partial differential equation solution u are bounded, 5.5.1 indicates that u can be approximated by trial functions to $O(h_{\text{max}}^2)$ accuracy, and thus the Galerkin error $\|U - u\|_\infty$ is expected to be $O(h_{\text{max}}^2)$ also.

The function $\Phi_i(x, y)$ is identically zero in all triangles except those that have (x_i, y_i) as a vertex, because in any other triangle, Φ_i is zero at all three vertices and therefore identically zero throughout the triangle, because of its linearity. Thus A_{ki} in 5.1.5b is zero unless the vertices k and i share a common triangle. If there are many triangles, then most of the A_{ki} will be zero, so the coefficient matrix of the linear system that must be solved to solve the partial differential equation 5.1.1 is sparse. (If the partial differential equation is nonlinear, the corresponding Jacobian matrix is sparse.) As discussed in Section 0.5, however, sparsity is not enough to ensure efficient solution of a linear system by using Gauss elimination. If a band solver is used, for example, the vertices (and thus the Φ_i) must be numbered in such a way that $A_{ki} \neq 0$ only when k and i are close to each other, that is, such that vertices that share a common triangle have numbers close together.

For a uniform triangulation of a rectangle, such as seen in Figure 5.5.2, such a numbering is easy to construct. The vertices in this figure may be numbered in "typewriter" order, just as is done for finite difference grids. For the $N \times M$ grid shown, the number of unknowns is NM, and the "half-bandwidth" ($\max_{A_{ki} \neq 0} |i - k|$), is $N + 1$. So the amount of work done by the band solver is about $NM(N + 1)^2$ multiplications, as shown in Section 0.4, or about $N^3 M$ (N^4 if $N = M$). This is comparable to the work for finite difference schemes.

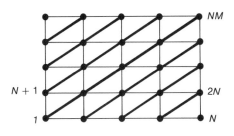

Figure 5.5.2

For an arbitrary triangulation of an arbitrary region, a good ordering for the unknowns is not as easy to determine. A popular scheme is the "reverse Cuthill-McKee" ordering, which proceeds as follows (see Figure 5.5.3): One vertex (usually a corner) is numbered 1. All of its "neighbors" (vertices with which it shares a common triangle) are given consecutive numbers 2 through, say, J. Next all as-yet-unnumbered neighbors of vertex 2 are numbered, beginning with $J + 1$, then all neighbors of vertex 3, and so on until all vertices are numbered. Normally this ordering is then reversed, even though

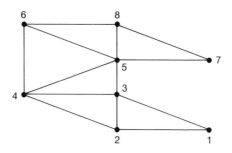

Figure 5.5.3
Cuthill–McKee Ordering

this does not affect the bandwidth, for reasons that are beyond the scope of this discussion.

Once the trial functions $\Omega, \Phi_1, \ldots, \Phi_M$ have been defined and their ordering fixed, all that remains is to calculate the elements of A and $\mathbf{b}$ from (the two-dimensional versions of) 5.1.5a–b, and to solve the linear system $A\mathbf{a} = \mathbf{b}$ for the unknown coefficients $a_i = U(x_i, y_i)$, by using a band solver or some other direct or iterative method. Numerical quadrature must generally be used, naturally, to compute the integrals in 5.1.5. From the discussion in Section 5.3, since the degree of the piecewise polynomials used is $n = 1$, the integration rules used must be exact for polynomials of degree $2n - 2 = 0$, that is, for constants. For the area integrals, the midpoint rule, which approximates an integral over a triangle by the triangle area times the value of the integrand at the center of gravity, is exact for linear polynomials and is therefore more than accurate enough. The boundary integrals may be evaluated by using a midpoint rule also.

5.6   Examples Using Triangular Elements

To illustrate the application of the finite element method to two-dimensional problems, let us solve

$$u_{xx} + u_{yy} + 1 = 0 \qquad \text{in } R$$

with $u = 0$ on the boundary.

Figure 5.6.1 shows a FORTRAN77 program that solves a general two-dimensional linear problem of the form 5.1.1, with essential boundary

```
      IMPLICIT DOUBLE PRECISION (A-H,O-Z)
C                               NTRI = NUMBER OF TRIANGLES
C                               M = NUMBER OF UNKNOWNS
      PARAMETER (NTRI=16, M=5)
      DIMENSION TRIX(3,NTRI),TRIY(3,NTRI),NODES(3,NTRI),AMAT(M,2*M),
     &          B(M),A(M)
      DIMENSION PHI(3),PHIX(3),PHIY(3)
C                               STATEMENT FUNCTIONS DEFINING PDE COEFFICIENTS
      FUND(X,Y) = 1.0
      FUNA(X,Y) = 0.0
      FUNF(X,Y) = 1.0
      FUNR(X,Y) = 0.0
C                               ARRAYS TRIX,TRIY HOLD THE VERTICES OF THE
C                               TRIANGLES. THE THREE VERTICES OF TRIANGLE
C                               KTRI ARE
C                                   (TRIX(J,KTRI), TRIY(J,KTRI))
C                               FOR J=1,2,3.
C
C                               ARRAY NODES HOLDS THE NUMBERS OF THE VERTICES.
C                               THE NUMBER OF THE VERTEX AT
C                                   (TRIX(J,KTRI), TRIY(J,KTRI))
C                               IS NODES(J,KTRI). THE NUMBER OF ANY VERTEX
C                               ON THE BOUNDARY IS TAKEN TO BE ZERO.
      TRIX(1,1) = 0.0
      TRIY(1,1) = 0.0
      NODES(1,1) = 0
      TRIX(2,1) = 0.0
      TRIY(2,1) =-1.0
      NODES(2,1) = 0
      TRIX(3,1) = 0.5
      TRIY(3,1) =-0.5
      NODES(3,1) = 1
         .
         .
         .
C                               CALCULATE HALF-BANDWIDTH, L
      L = 0
      DO 20 KTRI=1,NTRI
         DO 15 KX=1,3
            K = NODES(KX,KTRI)
            IF (K.GT.0) THEN
               DO 10 IX=1,3
                  I = NODES(IX,KTRI)
                  IF (I.GT.0) L = MAX(L,ABS(I-K))
   10          CONTINUE
            ENDIF
   15    CONTINUE
   20 CONTINUE
C                               ZERO BANDED COEFFICIENT MATRIX AMAT AND
```

(continued)

```
      C                           RIGHT-HAND SIDE VECTOR B.
            DO 30 K=1,M
               B(K) = 0.0
               DO 25 ISUB=1,2*L+1
                  AMAT(K,ISUB) = 0.0
         25    CONTINUE
         30 CONTINUE
      C                           BEGIN ELEMENT-BY-ELEMENT ASSEMBLY OF MATRIX
      C                           AND RIGHT-HAND SIDE VECTOR.
            DO 50 KTRI=1,NTRI
               X1 = TRIX(1,KTRI)
               Y1 = TRIY(1,KTRI)
               X2 = TRIX(2,KTRI)
               Y2 = TRIY(2,KTRI)
               X3 = TRIX(3,KTRI)
               Y3 = TRIY(3,KTRI)
               DET = (X2-X1)*(Y3-Y1) - (X3-X1)*(Y2-Y1)
      C                           PHI(J),PHIX(J),PHIY(J) REPRESENT THE VALUE,
      C                           X-DERIVATIVE AND Y-DERIVATIVE OF PHI-J AT
      C                           THE TRIANGLE MIDPOINT.
               PHI (1) = 1.0/3.0D0
               PHIX(1) = (Y2-Y3)/DET
               PHIY(1) = (X3-X2)/DET
               PHI (2) = 1.0/3.0D0
               PHIX(2) = (Y3-Y1)/DET
               PHIY(2) = (X1-X3)/DET
               PHI (3) = 1.0/3.0D0
               PHIX(3) = (Y1-Y2)/DET
               PHIY(3) = (X2-X1)/DET
      C                           OMEGA,OMEGAX,OMEGAY REPRESENT THE VALUE,
      C                           X-DERIVATIVE AND Y-DERIVATIVE OF OMEGA AT
      C                           THE TRIANGLE MIDPOINT.
               OMEGA  = 0.0
               OMEGAX = 0.0
               OMEGAY = 0.0
               DO 35 J=1,3
                  IF (NODES(J,KTRI).LE.0) THEN
                     R = FUNR(TRIX(J,KTRI),TRIY(J,KTRI))
                     OMEGA  = OMEGA  + R*PHI (J)
                     OMEGAX = OMEGAX + R*PHIX(J)
                     OMEGAY = OMEGAY + R*PHIY(J)
                  ENDIF
         35    CONTINUE
      C                           DMID,AMID,FMID ARE THE VALUES OF THE
      C                           COEFFICIENT FUNCTIONS D(X,Y), A(X,Y), F(X,Y)
      C                           AT THE TRIANGLE MIDPOINT.
               XMID = (X1+X2+X3)/3.0
               YMID = (Y1+Y2+Y3)/3.0
               DMID = FUND(XMID,YMID)
```

```
                 AMID = FUNA(XMID,YMID)
                 FMID = FUNF(XMID,YMID)
                 AREA = ABS(DET)
C                                      CALCULATE CONTRIBUTIONS OF TRIANGLE KTRI TO
C                                      MATRIX AND RIGHT-HAND SIDE INTEGRALS, USING
C                                      THE MIDPOINT RULE TO APPROXIMATE ALL
C                                      INTEGRALS.
            DO 45 KX=1,3
               K = NODES(KX,KTRI)
               IF (K.GT.0) THEN
                  B(K) = B(K) +
     &                (FMID*PHI(KX)
     &                -DMID*(OMEGAX*PHIX(KX) + OMEGAY*PHIY(KX))
     &                -AMID*OMEGA*PHI(KX)) *AREA
                  DO 40 IX=1,3
                     I = NODES(IX,KTRI)
                     IF (I.GT.0) THEN
                        ISUB = I-K+L+1
                        AMAT(K,ISUB) = AMAT(K,ISUB) +
     &                      (DMID*(PHIX(KX)*PHIX(IX) + PHIY(KX)*PHIY(IX))
     &                      +AMID*PHI(KX)*PHI(IX)) *AREA
                     ENDIF
40                CONTINUE
               ENDIF
45          CONTINUE
50       CONTINUE
C                                      CALL LBAND TO SOLVE BANDED LINEAR SYSTEM
         CALL LBAND(AMAT,A,B,M,M,L)
         PRINT 55, (A(I),I=1,M)
55       FORMAT (5F10.5)
         STOP
         END
```

Figure 5.6.1
FORTRAN77 Program Using Linear Triangular Elements

conditions, by using linear triangular elements. This program can be used to solve problems in any polygonal region, if the arrays TRIX, TRIY, and NODES are appropriately defined.

For our problem, $D(x, y) = 1$, $a(x, y) = 0$, $f(x, y) = 1$, and $r(x, y) = 0$. Formulas 5.1.5a–b are used to calculate the components of the coefficient matrix A and the right-hand side vector $\mathbf{b}$. Notice that A and $\mathbf{b}$ are assembled "element-by-element," that is, the contributions to all required integrals from one triangle are calculated before moving on to the next triangle. The area integrals in 5.1.5 are calculated by using the midpoint rule. The vertices are assigned numbers through the NODES array, and it is assumed that they are numbered in a way that will give a reasonably small bandwidth. The band solver LBAND of Figure 0.4.3 is called to solve the linear system $A\mathbf{a} = \mathbf{b}$ for the coefficients of the piecewise linear basis functions, which are the values of the approximate solution at the corresponding vertices.

When a triangulation consisting of 16 triangles was used, the solution at the point (0.5, 0.5) was calculated as 0.1346. A more accurate value, obtained by using PDE/PROTRAN (in a later example), is 0.1310.

Naturally, higher-order piecewise polynomial trial functions can be used to obtain higher-order accuracy. Programs that employ high-order elements are very complicated, however, so if high-order accuracy is desired, it is recommended that an existing finite element package be used, rather than homemade software.

One commercially available code is IMSL's PDE/PROTRAN [Sewell, 1985], which uses triangular elements with piecewise polynomials of degree 2,

```
$       PDE2D
C*** PDE/PROTRAN SOLVES  DA/DX + DB/DY + F = 0
        F = 1.0
        A = UX
        B = UY
        SYMMETRIC
C*** INITIAL TRIANGULATION SPECIFICATION
        VERTICES = (0,0) (0,-1) (1,-1) (1,0) (1,1) (0,1) (-1,1)
      * (-1,0) (-.5,.5) (.5,.5) (.5,-.5)
        TRIANGLES = (1,2,11,-1) (2,3,11,-1) (3,4,11,-1) (4,1,11,0)
      *             (4,5,10,-1) (5,6,10,-1) (6,1,10,0) (1,4,10,0)
      *             (6,7,9,-1) (7,8,9,-1) (8,1,9,-1) (1,6,9,0)
C*** NUMBER OF TRIANGLES AND GRADING DESIRED FOR FINAL TRIANGULATION
        NTRIANGLES = 100
        TRIDENSITY = 1./(X*X+Y*Y)
C*** USE CUBIC ELEMENTS
        DEGREE = 3
C*** OUTPUT PARAMETERS
        SAVEFILE = PLOT
        GRIDPOINTS = 17,17
$       PLOTSURFACE
        TITLE = 'SURFACE PLOT OF SOLUTION'
$       END
```

Figure 5.6.2
PDE/PROTRAN Input Program

3, or 4 to solve quite general steady-state, time-dependent and eigenvalue partial differential equation systems in general two-dimensional regions. To solve the above problem by using PDE/PROTRAN (Edition 1), the input program in Figure 5.6.2 is all the user must supply.

A detailed description of the PDE/PROTRAN input parameters is given in the PDE/PROTRAN User's Manual, and in Appendix 3 of Sewell, 1985.

Figure 5.6.3 shows the surface plot of the solution generated by this program. Piecewise cubic polynomials on the triangulation shown in Figure 5.6.4 were used as trial functions, and the linear system was solved by using a band solver, with a reverse Cuthill-McKee ordering of the unknowns. Since the solution is known to have singularities in its derivatives near the origin, a triangulation that is more refined near the origin is used.

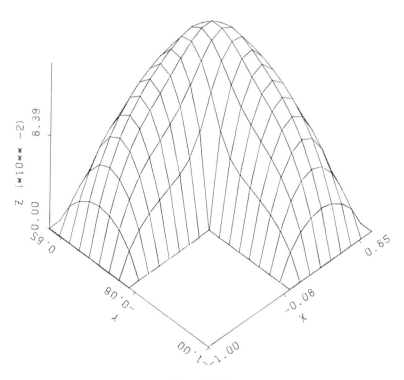

Figure 5.6.3
Surface Plot of Solution

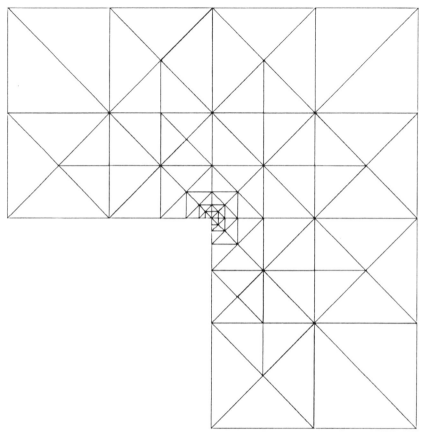

Figure 5.6.4
Triangulation Generated by PDE/PROTRAN

In fact, 5.5.1 can be generalized for interpolation using piecewise polynomials of degree n as follows:

$$\| W - u \|_\infty \leq M_n \max_k [h_k^{n+1}(D_k^{n+1}u)].$$

Since here the fourth derivatives, and thus $D_k^4 u$, become infinite as the origin is approached, the above formula suggests that the triangle size h_k be made small where $D_k^4 u$ is large. This is justified more thoroughly in Sewell, 1985, Section 2.4.

When the region has a curved boundary, normal (straight) triangles may be used to approximate the region if piecewise linear elements are used. But if this is done using n^{th} degree ($n > 1$) elements, the curvature of the boundary ruins the h_{max}^{n+1} accuracy normally expected. Then "isoparametric" elements— triangles with one edge that is a polynomial interpolant of degree n to the boundary curve—are used by PDE/PROTRAN to recover the optimal accuracy.[1]

Another software package that is widely used to solve linear, steady-state, partial differential equations in general two-dimensional regions (and rectangular three-dimensional regions) is ELLPACK [Rice and Boisvert, 1985]. ELLPACK offers a wide range of options for discretizing the partial differential equation, ordering the unknowns, and solving the resulting linear system. Both Galerkin and collocation finite element methods are available, as well as several finite difference discretization methods, and numerous direct and iterative linear system solvers are included.

Like PDE/PROTRAN, ELLPACK allows the user to specify a problem in a very high-level language and provides graphical output in various forms. The input interpreter (preprocessor) and the general framework for ELL-PACK are primarily the work of John Rice and coworkers at Purdue University, but the modules that implement the various discretization, ordering, and linear equation solution algorithms were contributed by experts from many universities and laboratories. ELLPACK is ideally suited for comparing competing algorithms for a particular class of partial differential equations.

5.7 Time-Dependent Problems

Many of the advantages of the finite element method, such as its ability to accurately represent solutions in general multi-dimensional domains, are still important when the problem is time-dependent. The finite element method is

---

[1] PDE/PROTRAN is available from:

> IMSL, Inc.
> 2500 ParkWest Tower One
> 2500 CityWest Blvd.
> Houston, TX 77042.

INTERPDE, a self-documenting interactive program that generates the input for PDE/ PROTRAN, is available from the author.

therefore widely used to discretize the spatial variables in time-dependent problems. Consider, for example, the time-dependent problem

$$c(x, y, z, t)u_t = \mathbf{V}^T(D(x, y, z, t)\mathbf{V}u) - a(x, y, z, t)u + f(x, y, z, t) \qquad \text{in } R,$$

$$u = r(x, y, z, t) \qquad \text{on part of the boundary } (\partial R_1),$$

$$D\partial u/\partial n = -p(x, y, z, t)u + q(x, y, z, t) \qquad \text{on the other part } (\partial R_2),$$

$$u = h(x, y, z) \qquad \text{at } t = 0. \tag{5.7.1}$$

It is assumed that $c > 0$ and $D \geq 0$.

In a manner analogous to what was done for the corresponding steady-state problem 5.1.1, we find the weak formulation of 5.7.1 by multiplying the partial differential equation and second boundary condition by a smooth function $\Phi(x, y, z)$ that satisfies $\Phi = 0$ on ∂R_1 and by integrating over R and ∂R_2:

$$\iiint\limits_R cu_t\Phi \, dx \, dy \, dz = \iiint\limits_R [\mathbf{V}^T(D\mathbf{V}u) - au + f]\Phi \quad dx \, dy \, dz$$

$$+ \iint\limits_{\partial R_2} \left[-D\frac{\partial u}{\partial n} - pu + q \right]\Phi \, dA.$$

Integrating by parts, remembering that $\Phi = 0$ on ∂R_1, gives (cf. 5.1.2)

$$\iiint\limits_R cu_t\Phi \, dx \, dy \, dz = \iiint\limits_R [-D\mathbf{V}u^T\mathbf{V}\Phi - au\Phi + f\Phi] \, dx \, dy \, dz$$

$$+ \iint\limits_{\partial R_2} [-pu\Phi + q\Phi] \, dA. \tag{5.7.2}$$

If u is *smooth* and satisfies 5.7.2, for all t, for arbitrary smooth $\Phi(x, y, z)$ vanishing on ∂R_1, then the steps from 5.7.1 to 5.7.2 are reversible, and therefore u satisfies the partial differential equation along with the second boundary condition. Thus, if it is required that $u(x, y, z, t)$ satisfy the first boundary condition (on ∂R_1) and the initial condition, in addition to being a smooth solution to the weak formulation 5.7.2, u will be a solution to the partial differential equation problem 5.7.1.

In the continuous-time Galerkin method, we attempt to find a solution to the weak formulation 5.7.2 of the form

$$U(x, y, z, t) = \Omega(x, y, z, t) + \sum_{i=1}^{M} a_i(t)\Phi_i(x, y, z), \tag{5.7.3}$$

where $\{\Phi_1, \ldots, \Phi_M\}$ is a set of linearly independent functions that vanish on ∂R_1 and Ω is another function that satisfies the essential boundary condition $\Omega = r$ on ∂R_1. Clearly, U will satisfy the essential boundary condition regardless of how the coefficients $a_i(t)$ are chosen.

As in the steady-state case, it is not possible to find coefficients such that U satisfies 5.7.2 for arbitrary Φ vanishing on ∂R_1, so it is only required that 5.7.2 be satisfied for $\Phi = \Phi_1, \ldots, \Phi_M$:

$$\iiint_R cU_t\Phi_k \, dx \, dy \, dz = \iiint_R [-D\mathbf{V}U^T\mathbf{V}\Phi_k - aU\Phi_k + f\Phi_k] \, dx \, dy \, dz$$

$$+ \iint_{\partial R_2} [-pU\Phi_k + q\Phi_k] \, dA \qquad (5.7.4)$$

Substituting 5.7.3 for U into 5.7.4 gives

$$\sum_{i=1}^{M} B_{ki}(t)a_i'(t) = -\sum_{i=1}^{M} A_{ki}(t)a_i(t) + b_k(t),$$

where

$$B_{ki}(t) = \iiint_R c\Phi_k\Phi_i \, dx \, dy \, dz, \qquad (5.7.5a)$$

$$A_{ki}(t) = \iiint_R [D\mathbf{V}\Phi_k^T\mathbf{V}\Phi_i + a\Phi_k\Phi_i] \, dx \, dy \, dz + \iint_{\partial R_2} p\Phi_k\Phi_i \, dA, \quad (5.7.5b)$$

$$b_k(t) = \iiint_R [-c\Omega_t\Phi_k - D\mathbf{V}\Omega^T\mathbf{V}\Phi_k - a\Omega\Phi_k + f\Phi_k] \, dx \, dy \, dz \quad (5.7.5c)$$

$$+ \iint_{R_2} [-p\Omega\Phi_k + q\Phi_k] \, dA.$$

In the steady-state case, the Galerkin method led to a system of algebraic equations for the unknown coefficients a_i. Here it leads to a system of ordinary differential equations

$$B(t)\mathbf{a}' = -A(t)\mathbf{a} + \mathbf{b}(t) \qquad (5.7.6)$$

for the unknown coefficient functions $a_i(t)$.

The initial values for this ordinary differential equation system are obtained by requiring that, at $t = 0$, U approximately satisfy the initial condition in 5.7.1,

$$U(x, y, z, 0) = \Omega(x, y, z, 0) + \Sigma \, a_i(0)\Phi_i(x, y, z) \approx h(x, y, z).$$

This is most easily accomplished by interpolation, that is, by requiring that the M numbers $a_i(0)$ be chosen so that $U(x, y, z, 0) = h(x, y, z)$ holds at M given points, but a projection may also be used, with $U(x, y, z, 0) - h(x, y, z)$ required to be orthogonal to each of the M functions Φ_k.

If it is assumed that a and p are nonnegative as well as D, then the matrix $A(t)$ is positive semidefinite, that is, $\mathbf{z}^T A \mathbf{z} \geq 0$ for any vector $\mathbf{z}$. Since A is defined the same for the time dependent case as in 5.1.5b, the proof following 5.1.5 that A is positive-definite is still valid, except that now D is only assumed nonnegative, not positive, and it can no longer be guaranteed that $\mathbf{z}^T A \mathbf{z}$ is strictly positive when $\mathbf{z} \neq \mathbf{0}$. The matrix $B(t)$, on the other hand, is easily shown to be positive-definite, since

$$\mathbf{z}^T B \mathbf{z} = \iiint\limits_{R} cZ^2 \, dx \, dy \, dz \geq 0,$$

where (as before) $Z(x, y, z) = \Sigma z_i \Phi_i(x, y, z)$. Equality holds above only (since $c > 0$) if Z, and therefore $\mathbf{z} = (z_1 \ldots z_M)$, is identically zero.

Since B is positive-definite, and therefore nonsingular, the ordinary differential equation system 5.7.6 can, in theory, be written in the standard form

$$\mathbf{a}' = -B^{-1}(t)A(t)\mathbf{a} + B^{-1}(t)\mathbf{b}(t). \tag{5.7.7}$$

Any of the initial value ordinary differential equation methods discussed in Chapter 1 could be applied to solve 5.7.7. However, since finite difference methods applied to equations similar to 5.7.1 lead to very stiff ordinary differential equation systems (see 2.1.8), we expect that 5.7.7 is stiff also, that is, that some eigenvalues of the Jacobian $-B^{-1}(t)A(t)$ are very large and negative. In fact, it is easy to see that the eigenvalues are all real and nonpositive, assuming $c > 0$, $D \geq 0$, $a \geq 0$, $p \geq 0$, for if

$$-B^{-1}A\mathbf{z} = \lambda \mathbf{z},$$

then

$$-\mathbf{z}^T A \mathbf{z} = \lambda \mathbf{z}^T B \mathbf{z},$$

which implies that $\lambda \leq 0$, since $\mathbf{z}^T A \mathbf{z} \geq 0$ and $\mathbf{z}^T B \mathbf{z} > 0$. It therefore seems reasonable to use the backward difference formulas (see Section 1.5) to solve

5.7.7, since they are well designed for stiff systems. These formulas are of the form

$$[\mathbf{a}(t_{k+1}) + \alpha_1\mathbf{a}(t_k) + \dots + \alpha_m\mathbf{a}(t_{k+1-m})]/\Delta t$$
$$= \beta_0[-B^{-1}(t_{k+1})A(t_{k+1})\mathbf{a}(t_{k+1}) + B^{-1}(t_{k+1})\mathbf{b}(t_{k+1})],$$

where values for $\alpha_1, \dots, \alpha_m$, $\beta_0 > 0$ are given in Table 1.5.1.

In this form, however, the backward difference methods are impractical, because, even though B is sparse, B^{-1} will generally be a full matrix. Thus they should be used in the form

$$B(t_{k+1})\mathbf{a}(t_{k+1}) + \alpha_1 B(t_{k+1})\mathbf{a}(t_k) + \dots + \alpha_m B(t_{k+1})\mathbf{a}(t_{k+1-m})$$
$$= -\Delta t\,\beta_0 A(t_{k+1})\mathbf{a}(t_{k+1}) + \Delta t\beta_0\mathbf{b}(t_{k+1}), \qquad (5.7.8)$$

This implicit method requires the solution each time step of a linear system with matrix $B(t_{k+1}) + \Delta t\,\beta_0 A(t_{k+1})$, to find the unknown $\mathbf{a}(t_{k+1})$. With any reasonable choice of trial functions $\Phi_1, \dots, \Phi_M$, this matrix is sparse, and positive-definite for sufficiently small Δt (since B is positive-definite, and A is symmetric). In fact, if $a \geq 0$ and $p \geq 0$, it is positive-definite for any Δt, since then A is positive-semidefinite, and $\beta_0 > 0$.

5.8   A One-Dimensional Example

As an example of the use of the continuous-time Galerkin method to solve time-dependent problems, consider the diffusion problem 2.3.1, previously solved using finite differences. This problem does not quite fit the format of 5.7.1, because of the nonsymmetric first-order term, xtu_x. However, an expression of the form $u_{xx} + b(x, t)u_x$ can be put into symmetric form by multiplying through by e^B, where $B_x = b$:

$$e^{B(x,\,t)}[u_{xx} + b(x, t)u_x] = [e^{B(x,\,t)}u_x]_x.$$

In the case of problem 2.3.1, $b = xt$, so $B = \frac{1}{2}x^2 t$, and this problem can be rewritten in a form consistent with 5.7.1:

$$e^{(1/2)x^2 t}u_t = [e^{(1/2)x^2 t}u_x]_x + e^{(1/2)x^2 t}xtu,$$

$$u(0, t) = e^t, \qquad 0 \leq t \leq 1, \qquad\qquad (5.8.1)$$

$$e^{(1/2)t}u_x(1, t) = -e^{(1/2)t}u(1, t),$$

$$u(x, 0) = e^{-x}, \qquad 0 \leq x \leq 1.$$

This problem has the exact solution $u = e^{t-x}$.

If it were not possible to manipulate this problem into the form 5.7.1, the finite element method could still be used. A new weak formulation could be recomputed, as was done for the nonstandard boundary value problem of Section 5.3, and the approximate solution would be required to satisfy it for $\Phi = \Phi_k$.

The coefficients in 5.7.1 corresponding to the problem 5.8.1 are then

$$c(x, t) = e^{(1/2)x^2 t}, \qquad D(x, t) = e^{(1/2)x^2 t},$$

$$a(x, t) = -e^{(1/2)x^2 t} xt, \qquad f(x, t) = 0,$$

$$r(t) = e^t, \qquad p(t) = e^{(1/2)t},$$

$$q(t) = 0, \qquad h(x) = e^{-x}.$$

Formulas (5.7.5) can be used to determine the coefficients in the ordinary differential equation system $B(t)\mathbf{a}' = -A(t)\mathbf{a} + \mathbf{b}(t)$, corresponding to 5.8.1:

$$B_{ki}(t) = \int_0^1 e^{(1/2)x^2 t} \Phi_k \Phi_i \, dx,$$

$$A_{ki}(t) = \int_0^1 e^{(1/2)x^2 t} [\Phi_k' \Phi_i' - xt\Phi_k \Phi_i] \, dx + e^{(1/2)t} \Phi_k(1)\Phi_i(1), \qquad (5.8.2)$$

$$b_k(t) = \int_0^1 e^{(1/2)x^2 t} [-\Omega_t \Phi_k - \Omega_x \Phi_k' + xt\Omega \Phi_k] \, dx - e^{(1/2)t}\Omega(1, t)\Phi_k(1).$$

The trial functions $\Omega, \Phi_1, \ldots, \Phi_M$ will again be chosen from the cubic Hermite polynomials defined in 5.3.5:

$$\Omega(x, t) = e^t H_0(x),$$

$$\{\Phi_1, \ldots, \Phi_{2N+1}\} = \{S_0, H_1, S_1, \ldots, H_N, S_N\}.$$

Thus the number of unknown coefficients in 5.7.3 is $M = 2N + 1$.

Using the properties 5.3.6, it is easy to verify that $\Omega(0, t) = e^t$ and that $\Phi_k(0) = 0$, so that these are acceptable basis functions, and that the approximate solution $U(x, t)$ given by (see 5.7.3)

$$U(x, t) = \Omega(x, t) + \sum_{i=1}^{2N+1} a_i(t)\Phi_i(x)$$

$$= e^t H_0(x) + \sum_{i=0}^{N} a_{2i+1}(t)S_i(x) + \sum_{i=1}^{N} a_{2i}(t)H_i(x)$$

satisfies the essential boundary condition (at $x = 0$), regardless of the way in which the unknown coefficients $a_i(t)$ are chosen.

From the properties 5.3.6, it is also clear that

$$U(x_j, t) = a_{2j}(t), \qquad j = 1, \ldots, N,$$

$$U_x(x_j, t) = a_{2j+1}(t), \qquad j = 0, \ldots, N.$$

Thus if it is required that U and U_x interpolate to the initial condition and its derivative, this provides initial conditions for the unknown coefficients:

$$a_{2j}(0) = U(x_j, 0) = e^{-x_j}, \qquad j = 1, \ldots, N,$$

$$a_{2j+1}(0) = U_x(x_j, 0) = -e^{-x_j}, \qquad j = 0, \ldots, N.$$

(5.8.3)

So now the ordinary differential equation problem 5.7.6, with coefficients given by 5.8.2 and with initial conditions given by 5.8.3, must be solved to provide a solution to the time-dependent problem 5.8.1. This was done using the fourth-order backward difference method, in the form 5.7.8, with coefficients $\beta_0, \alpha_1, \ldots, \alpha_4$ given by Table 1.5.1, with $m = 4$. Exact starting values were used, and all integrals in 5.8.2 were calculated using the 3-point Gaussian quadrature formula, as in Section 5.3. The matrix $B(t_{k+1}) + \Delta t \beta_0 A(t_{k+1})$ in the linear system which must be solved at each time step is banded, with half-bandwidth equal to 3 (cf. Figure 5.3.2).

The errors are reported in Table 5.8.1. In the first two experiments, the (constant) space stepsize was made small enough so that the error is primarily due to the time discretization. By comparing the two maximum errors ("maximum error" means maximum over 100 uniformly spaced points) at $t = 1$, it can be seen that decreasing the time stepsize (Δt) by a factor of 2 decreased the error by a factor of 14.3, approximately confirming the $O(\Delta t^4)$ error behavior of the fourth-order backward difference formula. In the last two experiments, the time stepsize was made small enough so that the error is primarily due to the (finite element) space discretization. By comparing the two errors at $t = 1$, it can be seen that decreasing the space stepsize (Δx) by a factor of 2 decreased the error by a factor of 13.7, which suggests that the spatial discretization error is the same here as for steady-state problems using cubic Hermite basis functions, namely $O(\Delta x^4)$ (cf. 5.3.7).

When Euler's method is applied to solve this same problem, the results confirm the very stiff nature of this ordinary differential equation system, and the time stepsize had to be very small to avoid instability (Table 5.8.2). With a system of the form 5.7.6, with $B \neq I$, even Euler's method is implicit, and a linear system must be solved each step. Thus Euler and other nonstiff

Table 5.8.1

Maximum Errors, Using Fourth-Order Backward Difference Formula

Δx	Δt	t	Maximum Error
0.0100	0.0500	0.1	(Specified Exactly)
		0.2	$0.208E - 07$
		0.3	$0.912E - 07$
		0.4	$0.146E - 06$
		0.5	$0.192E - 06$
		0.6	$0.236E - 06$
		0.7	$0.279E - 06$
		0.8	$0.322E - 06$
		0.9	$0.367E - 06$
		1.0	$0.416E - 06$
0.0100	0.1000	0.1	(Specified Exactly)
		0.2	(Specified Exactly)
		0.3	(Specified Exactly)
		0.4	$0.654E - 06$
		0.5	$0.174E - 05$
		0.6	$0.282E - 05$
		0.7	$0.370E - 05$
		0.8	$0.444E - 05$
		0.9	$0.516E - 05$
		1.0	$0.594E - 05$
0.1000	0.0050	0.1	$0.146E - 06$
		0.2	$0.137E - 06$
		0.3	$0.142E - 06$
		0.4	$0.157E - 06$
		0.5	$0.177E - 06$
		0.6	$0.198E - 06$
		0.7	$0.220E - 06$
		0.8	$0.245E - 06$
		0.9	$0.272E - 06$
		1.0	$0.302E - 06$
0.2000	0.0050	0.1	$0.228E - 05$
		0.2	$0.207E - 05$
		0.3	$0.211E - 05$
		0.4	$0.222E - 05$
		0.5	$0.238E - 05$
		0.6	$0.266E - 05$
		0.7	$0.298E - 05$
		0.8	$0.333E - 05$
		0.9	$0.371E - 05$
		1.0	$0.413E - 05$

Table 5.8.2
Maximum Errors, Using Euler's Method

Δx	Δt	t	Maximum Error
0.2000	0.0002	0.005	0.336E − 05
		0.010	0.366E − 05
		0.015	0.389E − 05
		0.020	0.409E − 05
		0.025	0.428E − 05
		0.030	0.447E − 05
		0.035	0.465E − 05
		0.040	0.484E − 05
		0.045	0.508E − 05
		0.050	0.532E − 05
0.2000	0.0010	0.005	0.969E − 04
		0.010	0.110E − 01
		0.015	0.127E + 01
		0.020	0.148E + 03
		0.025	0.173E + 05
		0.030	0.202E + 07
		0.035	0.237E + 09
		0.040	0.280E + 11
		0.045	0.330E + 13
		0.050	0.392E + 15

methods that are explicit when applied to the ordinary differential equation systems (e.g., 2.1.8) arising from finite difference spatial discretizations no longer have the advantage of explicitness and thus have little to offer here.

Comparison with Table 2.3.1 shows that, when using the second-order finite difference spatial discretization, a much smaller space stepsize (Δx) is required for comparable accuracy, compared with the fourth-order Galerkin discretization.

If c, D, a, and p are independent of time, then by 5.7.5 the matrices B and A are also time-independent. If Δt is also constant, the matrix $B + \Delta t \beta_0 A$, which must be inverted each step when a backward difference method is used to solve 5.7.6, does not change. This can be taken advantage of in that once its LU decomposition (see Section 0.3) has been found, each subsequent step involves only the solution of two triangular, banded systems, which is less work than re-solving the system (although the gain is not nearly as important here as for multi-dimensional problems, where the bandwidth is much larger). Even if the matrix is not constant, it may vary slowly, so that it does not need to be updated and refactored every step.

A good stiff ordinary differential equation solver may be invoked to solve 5.7.6, and it will presumably take care of varying the stepsize appropriately to control the error and updating the matrix at appropriate intervals. While most ordinary differential equation software packages are designed only to handle problems of the form 5.7.6 with $B = I$, some [Hindmarsh, 1984] are available for the more general problem. Reduction of a time-dependent partial differential equation to a system of ordinary differential equations using some spatial discretization technique, and using an ordinary differential equation solver on this system, is referred to as a "method of lines" solution.

IMSL subroutine MOLCH [Sewell, 1982] is typical of the "method of lines" software available to solve time-dependent partial differential equation systems. MOLCH solves time-dependent partial differential equations in one space variable by using a collocation finite element method to discretize the space variable. That is, it assumes an approximate solution $U(x, t)$ of the form 5.7.3, with Ω and Φ_i chosen from the cubic Hermite basis functions 5.3.5; and rather than requiring that the partial differential equation be satisfied when it is multiplied by $\Phi_k(x)$ and integrated, it simply requires that the partial differential equation be satisfied exactly at two selected points in each interval $[x_i, x_{i+1}]$. As in the Galerkin case, this results in a system of ordinary differential equations of the form 5.7.6 for the coefficient functions $a_i(t)$. This stiff ordinary differential equation system is solved using the backward difference methods of (a modified version of) IMSL routine IVPAG.

5.9   A Time-Dependent Example Using Triangular Elements

To illustrate the application of the finite element method to two-dimensional time-dependent problems, let us solve

$$u_t = u_{xx} + u_{yy} - (1 + x + y) \qquad \text{in } R$$

with

$$u = (1 - t)(1 + x + y) \qquad \text{on the boundary,}$$

and

$$u = 1 + x + y \qquad \text{at } t = 0.$$

Figure 5.9.1 shows a FORTRAN77 program which solves a general two-dimensional time-dependent linear problem of the form 5.7.1, with essential

```
      IMPLICIT DOUBLE PRECISION (A-H,O-Z)
C                             NTRI = NUMBER OF TRIANGLES
C                             M = NUMBER OF UNKNOWNS
C                             NSTEP = NUMBER OF TIME STEPS
      PARAMETER (NTRI=16, M=5, NSTEP=10)
      DIMENSION TRIX(3,NTRI),TRIY(3,NTRI),NODES(3,NTRI),AMAT(M,2*M),
     &          BMAT(M,2*M),A(M),B(M)
      DIMENSION PHI(3),PHIX(3),PHIY(3)
C                             STATEMENT FUNCTIONS DEFINING PDE COEFFICIENTS
      FUNC(X,Y,T) = 1.0
      FUND(X,Y,T) = 1.0
      FUNA(X,Y,T) = 0.0
      FUNF(X,Y,T) = -(1.0+X+Y)
      FUNR(X,Y,T) = (1.0-T)*(1.0+X+Y)
      FUNRT(X,Y,T) = -(1.0+X+Y)
      FUNH(X,Y) = 1.0+X+Y
C                             TRIX,TRIY,NODES DEFINED AS IN FIGURE 5.6.1.
      TRIX(1,1) = 0.0
      TRIY(1,1) = 0.0
      NODES(1,1) = 0
      TRIX(2,1) = 0.0
      TRIY(2,1) =-1.0
      NODES(2,1) = 0
      TRIX(3,1) = 0.5
      TRIY(3,1) =-0.5
      NODES(3,1) = 1
         .
         .
         .
C                             CALCULATE HALF-BANDWIDTH, L
      L = 0
      DO 20 KTRI=1,NTRI
         DO 15 KX=1,3
            K = NODES(KX,KTRI)
            IF (K.GT.0) THEN
C                             ASSIGN INITIAL VALUES
               A(K) = FUNH(TRIX(KX,KTRI) , TRIY(KX,KTRI))
               DO 10 IX=1,3
                  I = NODES(IX,KTRI)
                  IF (I.GT.0) L = MAX(L,ABS(I-K))
   10          CONTINUE
            ENDIF
   15    CONTINUE
   20 CONTINUE
C                             TAKE NSTEP STEPS OF SIZE DT
      DT = 1.0D0/NSTEP
      DO 70 KT=1,NSTEP
      TKP1 = KT*DT
```

(*continued*)

```
C                                      ZERO BAND MATRICES AMAT,BMAT AND VECTOR B.
      DO 30 K=1,M
         B(K) = 0.0
         DO 25 ISUB=1,2*L+1
            AMAT(K,ISUB) = 0.0
            BMAT(K,ISUB) = 0.0
   25    CONTINUE
   30 CONTINUE
C                                      BEGIN ELEMENT-BY-ELEMENT ASSEMBLY OF AMAT,BMAT,B
      DO 50 KTRI=1,NTRI
         X1 = TRIX(1,KTRI)
         Y1 = TRIY(1,KTRI)
         X2 = TRIX(2,KTRI)
         Y2 = TRIY(2,KTRI)
         X3 = TRIX(3,KTRI)
         Y3 = TRIY(3,KTRI)
         DET = (X2-X1)*(Y3-Y1) - (X3-X1)*(Y2-Y1)
C                                      PHI(J),PHIX(J),PHIY(J) REPRESENT THE VALUE,
C                                      X-DERIVATIVE AND Y-DERIVATIVE OF PHI-J AT
C                                      THE TRIANGLE MIDPOINT.
         PHI (1) = 1.0/3.0D0
         PHIX(1) = (Y2-Y3)/DET
         PHIY(1) = (X3-X2)/DET
         PHI (2) = 1.0/3.0D0
         PHIX(2) = (Y3-Y1)/DET
         PHIY(2) = (X1-X3)/DET
         PHI (3) = 1.0/3.0D0
         PHIX(3) = (Y1-Y2)/DET
         PHIY(3) = (X2-X1)/DET
C                                      OMEGA,OMEGAT,OMEGAX,OMEGAY REPRESENT THE VALUE,
C                                      T-DERIVATIVE, X-DERIVATIVE, AND Y-DERIVATIVE OF
C                                      OMEGA AT THE TRIANGLE MIDPOINT.
         OMEGA  = 0.0
         OMEGAT = 0.0
         OMEGAX = 0.0
         OMEGAY = 0.0
         DO 35 J=1,3
            IF (NODES(J,KTRI).LE.0) THEN
               R  = FUNR(TRIX(J,KTRI),TRIY(J,KTRI),TKP1)
               RT = FUNRT(TRIX(J,KTRI),TRIY(J,KTRI),TKP1)
               OMEGA  = OMEGA  + R*PHI (J)
               OMEGAT = OMEGAT + RT*PHI(J)
               OMEGAX = OMEGAX + R*PHIX(J)
               OMEGAY = OMEGAY + R*PHIY(J)
            ENDIF
   35    CONTINUE
C                                      CMID,DMID,AMID,FMID ARE THE VALUES OF THE
C                                      COEFFICIENT FUNCTIONS C(X,Y,T), D(X,Y,T),
C                                      A(X,Y,T), F(X,Y,T) AT THE TRIANGLE MIDPOINT.
         XMID = (X1+X2+X3)/3.0
         YMID = (Y1+Y2+Y3)/3.0
         CMID = FUNC(XMID,YMID,TKP1)
         DMID = FUND(XMID,YMID,TKP1)
```

```
        AMID = FUNA(XMID,YMID,TKP1)
        FMID = FUNF(XMID,YMID,TKP1)
        AREA = ABS(DET)
C                              CALCULATE CONTRIBUTIONS OF TRIANGLE KTRI TO
C                              MATRICES AMAT,BMAT AND VECTOR B, USING
C                              THE MIDPOINT RULE TO APPROXIMATE ALL
C                              INTEGRALS.
        DO 45 KX=1,3
          K = NODES(KX,KTRI)
          IF (K.GT.0) THEN
            B(K) = B(K) +
     &        (-CMID*OMEGAT*PHI(KX)
     &        -DMID*(OMEGAX*PHIX(KX) + OMEGAY*PHIY(KX))
     &        -AMID*OMEGA*PHI(KX)
     &        +FMID*PHI(KX)) *AREA
            DO 40 IX=1,3
              I = NODES(IX,KTRI)
              IF (I.GT.0) THEN
                ISUB = I-K+L+1
                AMAT(K,ISUB) = AMAT(K,ISUB) +
     &                  (DMID*(PHIX(KX)*PHIX(IX) + PHIY(KX)*PHIY(IX))
     &                  +AMID*PHI(KX)*PHI(IX)) *AREA
                BMAT(K,ISUB) = BMAT(K,ISUB) +
     &                  CMID*PHI(KX)*PHI(IX) *AREA
              ENDIF
40          CONTINUE
          ENDIF
45      CONTINUE
50 CONTINUE
C                              BACKWARD DIFFERENCE METHOD REQUIRES
C                              SOLUTION OF
C                                (BMAT + DT*AMAT)*AKP1 = DT*B + BMAT*AK
C                              EACH STEP.
    DO 60 K=1,M
      B(K) = DT*B(K)
      DO 55 ISUB=1,2*L+1
        I = K+ISUB-L-1
        IF (I.GE.1.AND.I.LE.M) THEN
            AMAT(K,ISUB) = BMAT(K,ISUB) + DT*AMAT(K,ISUB)
            B(K) = B(K) + BMAT(K,ISUB)*A(I)
        ENDIF
55    CONTINUE
60 CONTINUE
C                          CALL LBAND TO SOLVE BANDED LINEAR SYSTEM
    CALL LBAND(AMAT,A,B,M,M,L)
    PRINT 65, TKP1,(A(I),I=1,M)
65 FORMAT (6F10.5)
70 CONTINUE
    STOP
    END
```

Figure 5.9.1
FORTRAN77 Program Using Linear Triangular Elements

boundary conditions, by using linear triangular elements to generate the
system 5.7.6 of ordinary differential equations, and by using the first-order
backward difference method (Table 1.5.1, $m = 1$) to solve this ordinary
differential equation system.

For our problem $c(x, y, t) = 1$, $D(x, y, t) = 1$, $a(x, y, t) = 0$ $f(x, y, t) =$
$-(1 + x + y)$, $r(x, y, t) = (1 - t)(1 + x + y)$, and $h(x, y) = 1 + x + y$. Formulas (5.7.5a–c) are used to calculate the coefficients of the matrices A and B,
and of the vector $\mathbf{b}$. The program is similar to the steady-state program of
Figure 5.6.1, but each time step the first-order backward difference method
requires that the linear system

$$[B(t_{k+1}) + \Delta t A(t_{k+1})]\mathbf{a}(t_{k+1}) = \Delta t \mathbf{b}(t_{k+1}) + B(t_{k+1})\mathbf{a}(t_k)$$

be solved (see 5.7.8). This system is solved, as before, by the band solver
LBAND of Figure 0.4.3.

The triangulation used was the same as in Figure 5.6.1, and a time step of
$\Delta t = 0.1$ was chosen. Because the exact solution $u = (1 - t)(1 + x + y)$ is
linear in x, y, and t, the observed error is exactly zero.

5.10   The Eigenvalue Problem

Let us see how to use the finite element method to solve the linear eigenvalue
problem

$$\mathbf{V}^T(D(x, y, z)\mathbf{V}u) - a(x, y, z)u = \lambda\rho(x, y, z)u \qquad \text{in } R, \qquad (5.10.1)$$

$$u = 0 \qquad\qquad\qquad \text{on part of the boundary of } R\ (\partial R_1),$$

$$D\partial u/\partial n = -p(x, y, z)u \qquad \text{on the other part } (\partial R_2).$$

It is assumed that $\rho > 0$. We want to find values of the eigenvalue λ for which
the homogeneous problem 5.10.1 has nonzero solutions (the eigenfunctions).

As usual, we multiply the partial differential equation and second boundary condition by a smooth function Φ, which is arbitrary except that it is
required to satisfy $\Phi = 0$ on ∂R_1. If these are integrated over R and ∂R_2,
respectively, then

$$\iiint\limits_R [\mathbf{V}^T(D\mathbf{V}u) - au]\Phi \, dx \, dy \, dz + \iint\limits_{\partial R_2} \left[-D\frac{\partial u}{\partial n} - pu \right]\Phi \, dA$$

$$= \lambda \iiint\limits_R \rho u\Phi \, dx \, dy \, dz.$$

Using $\Phi\mathbf{V}^T(D\mathbf{V}\mathbf{u}) = \mathbf{V}^T(\Phi D\mathbf{V}\mathbf{u}) - D\mathbf{V}\mathbf{u}^T\mathbf{V}\Phi$ and the divergence theorem,

$$\iiint\limits_{R} [-D\mathbf{V}\mathbf{u}^T\mathbf{V}\Phi - au\Phi]\, dx\, dy\, dz + \iint\limits_{\partial R} \Phi D\mathbf{V}\mathbf{u}^T\mathbf{n}\, dA$$

$$+ \iint\limits_{\partial R_2} [-\Phi D\mathbf{V}\mathbf{u}^T\mathbf{n} - pu\Phi]\, dA = \lambda \iiint\limits_{R} \rho u\Phi\, dx\, dy\, dz.$$

The integrand in the first boundary integral is nonzero only on ∂R_2, since $\Phi = 0$ on ∂R_1. Thus

$$\iiint\limits_{R} [-D\mathbf{V}\mathbf{u}^T\mathbf{V}\Phi - au\Phi]\, dx\, dy\, dz - \iint\limits_{\partial R_2} pu\Phi\, dA = \lambda \iiint\limits_{R} \rho u\Phi\, dx\, dy\, dz.$$

$$(5.10.2)$$

We attempt to find eigenfunctions for this "weak formulation," of the form

$$U(x, y, z) = \sum_{i=1}^{M} a_i\Phi_i(x, y, z), \qquad (5.10.3)$$

where $\{\Phi_1, \ldots, \Phi_M\}$ is a set of linearly independent functions that vanish on ∂R_1. Clearly, U will satisfy the essential boundary condition regardless of the values chosen for $a_1 \ldots a_M$.

As before, U is only required to satisfy (5.10.2) for $\Phi = \Phi_1, \ldots, \Phi_M$ (each of which vanishes on ∂R_1):

$$\iiint\limits_{R} [-D\mathbf{V}U^T\mathbf{V}\Phi_\mathbf{k} - aU\Phi_k]\, dx\, dy\, dz - \iint\limits_{\partial R_2} pU\Phi_k\, dA$$

$$= \lambda \iiint\limits_{R} \rho U\Phi_k\, dx\, dy\, dz,$$

which can be written as a generalized matrix eigenvalue problem of the form

$$A\mathbf{a} = \lambda B\mathbf{a}, \qquad (5.10.4)$$

where $\mathbf{a} = (a_1, \ldots, a_M)$, or

$$\sum_{i=1}^{M} A_{ki}a_i = \lambda \sum_{i=1}^{M} B_{ki}a_i,$$

where

$$A_{ki} = \iiint_R [-D\nabla\Phi_k^T\nabla\Phi_i - a\Phi_k\Phi_i]\, dx\, dy\, dz - \iint_{\partial R_2} p\Phi_k\Phi_i\, dA, \quad (5.10.5a)$$

$$B_{ki} = \iiint_R p\Phi_k\Phi_i\, dx\, dy\, dz. \qquad\qquad\qquad\qquad\qquad\qquad (5.10.5b)$$

From (5.10.5) it is clear that the matrices A and B are symmetric. B is also positive-definite, under the assumption that $\rho > 0$, because if $\mathbf{z}$ is an arbitrary vector with components $z_1, \dots, z_M$, then

$$\mathbf{z}^T B\mathbf{z} = \sum_{k=1}^{M}\left(\sum_{i=1}^{M} B_{ki}z_i\right)z_k = \iiint_R \rho Z^2\, dx\, dy\, dz \geq 0,$$

where $Z(x, y, z) \equiv \Sigma\, z_i\Phi_i(x, y, z)$. If equality holds ($\mathbf{z}^T B\mathbf{z} = 0$) then clearly (since $\rho > 0$), $Z \equiv 0$, and by the independence of the Φ_i, the z_i must all be zero. Hence $\mathbf{z} = \mathbf{0}$, and so B is positive-definite.

The generalized matrix eigenvalue problem 5.10.4, where A and B are symmetric and B is positive-definite, is identical to the problem 4.10.4 already studied in Sections 4.10 and 4.11. The matrix B is no longer diagonal (A and B will usually be banded), but the analysis in those sections was done assuming B to be a general positive-definite matrix, in anticipation of the finite element case.

If the Cholesky decomposition LL^T of B is formed, this problem has the same eigenvalues as the standard matrix eigenvalue problem 4.10.5, but since B (and thus L) are now not diagonal matrices, $H = L^{-1}AL^{-T}$ is generally a full matrix so that solving 4.10.5 is not feasible. Software does exist, however, which handles $A\mathbf{a} = \lambda B\mathbf{a}$ directly and efficiently when A and B are banded. Alternatively, the inverse power method (4.11.1) or the shifted inverse power method (4.11.6) may be used to find one eigenvalue-eigenvector pair at a time.

5.11   Eigenvalue Examples

As an example, let us solve again the one-dimensional problem (4.11.7) that, after multiplying by e^{-2x}, has the symmetric form

$$(-e^{-2x}u_x)_x - e^{-2x}u = \lambda e^{-2x}u, \qquad (5.11.1)$$

$$u(0) = 0,$$

$$-e^{-2}u_x(1) = 0.$$

This fits the format 5.10.1, with $-D(x) = a(x) = e^{-2x}$, $\rho(x) = e^{-2x}$ and $p = 0$. Thus the coefficients of the matrices A and B can be calculated by using formulas 5.10.5, with the understanding that R is the interval $[0, 1]$ and ∂R_2 is the point $x = 1$. Then

$$A_{ki} = \int_0^1 e^{-2x}[\Phi'_k\Phi'_i - \Phi_k\Phi_i]\, dx,$$

$$B_{ki} = \int_0^1 e^{-2x}\Phi_k\Phi_i\, dx.$$

The basis functions $\Phi_i(x)$ are chosen to be $\{\Phi_1, \ldots, \Phi_{2N+1}\} = \{S_0, H_1, S_1, \ldots, H_N, S_N\}$, where S_i and H_i are the cubic Hermite functions defined in formulas 5.3.5. Note that all of the Φ_i are zero on ∂R_1 (the point $x = 0$), as required, since $H_0(x)$, the only cubic Hermite basis function that is nonzero at $x = 0$, is not included in the set $\{\Phi_i\}$. The matrices A and B have a band structure similar to that of Figure 5.3.2, and their elements are calculated by using a 3-point Gaussian quadrature formula.

The matrix eigenvalue problem 5.10.4 was solved by using the shifted inverse power iteration 4.11.6. Results for various values of α and $\Delta x (\equiv 1/N)$ are shown in Table 5.11.1. Initial values of $U(x) = 1$ were used in all trials.

By comparing the errors with the two values of Δx, the experimental rates of convergence can be calculated. For the first and second eigenvalues, we calculate orders of $\ln(0.0002955/0.0000032)/\ln(2) = 6.5$ and $\ln(0.0397164/0.0008086)/\ln(2) = 5.6$, respectively. For the first and second eigenfunctions, we calculate orders of 3.4 and 3.3. These experimental orders agree fairly well with the theory [Varga, 1971], which says that the error in the eigenvalue and eigenfunction approximations should be $O(\Delta x^{2n})$ and $O(\Delta x^{n+1})$, respectively, where n is the degree of the piecewise polynomial trial function space, which is 3 in this case.

The FORTRAN77 program used to solve 5.11.1 is shown in Figure 5.11.1. The subprograms of Figure 5.3.3b and the band solver LBAND (Figure 0.4.3)

Table 5.11.1
Inverse Power Method Results

Δx	α	Number of iterations	μ_{final}	Error in eigenvalue	Maximum error in eigenfunction
0.5	0.0	6	4.1155629	0.0002955	0.0006842
	4.0	3	4.1155629	0.0002955	0.0006842
	20.0	9	24.0996256	0.0397164	0.0136242
0.25	0.0	6	4.1158552	0.0000032	0.0000633
	4.0	3	4.1158552	0.0000032	0.0000633
	20.0	9	24.1401506	0.0008086	0.0014286

```
      IMPLICIT DOUBLE PRECISION(A-H,O-Z)
      EXTERNAL FUNA,FUNB
      CHARACTER*1 HORS
C                                  NSUBS = NUMBER OF SUBINTERVALS
      PARAMETER (NSUBS=4)
C                                  BREAK POINTS ARE XPTS(K), K=0,N
      COMMON /BLKX/ N,XPTS(0:NSUBS)
C                                  UNKNOWNS ARE A(I), I=1,M
      COMMON /BLKA/ M,A(2*NSUBS+1)
      DIMENSION AMAT(2*NSUBS+1,7),F(2*NSUBS+1),AOLD(2*NSUBS+1)
      EXACT(X) = EXP(X)*SIN(2.0287579*X)
      N = NSUBS
      M = 2*NSUBS+1
      ALPHA = 4.0
      DO 5 K=0,N
         XPTS(K) = DBLE(K)/N
    5 CONTINUE
C                                  INITIAL VALUES ASSIGNED
      DO 10 I=1,M
         CALL LOCATE(I,HORS,K)
         IF (HORS.EQ.'H') A(I) = 1.0
         IF (HORS.EQ.'S') A(I) = 0.0
   10 CONTINUE
C                                  DO ITMAX INVERSE POWER ITERATIONS
      ITMAX = 15
      DO 45 ITER=1,ITMAX
C                                  SAVE PREVIOUS A VECTOR
         DO 15 I=1,M
            AOLD(I) = A(I)
   15    CONTINUE
C                                  CALCULATE MATRIX A-ALPHA*B, AND F = B*AOLD
         DO 25 K=1,M
            F(K) = 0.0
            DO 20 ISUB=1,7
               I = K-4+ISUB
               IF (I.GE.1.AND.I.LE.M) THEN
                  AMAT(K,ISUB) = GAUSS(FUNA,K,I)
                  BMAT = GAUSS(FUNB,K,I)
                  F(K) = F(K) + BMAT*A(I)
                  AMAT(K,ISUB) = AMAT(K,ISUB) - ALPHA*BMAT
               ENDIF
   20       CONTINUE
   25    CONTINUE
C                                  SOLVE LINEAR SYSTEM USING BAND SOLVER
         L = 3
         CALL LBAND(AMAT,A,F,M,M,L)
C                                  ESTIMATE EIGENVALUE
         ANUM = 0.0
         ADEN = 0.0
         DO 30 I=1,M
```

```
              ANUM = ANUM + F(I)*AOLD(I)
              ADEN = ADEN + F(I)*A(I)
    30    CONTINUE
          EIGEN = ANUM/ADEN + ALPHA
C                              CALCULATE MAXIMUM ERROR
          ERMAX = 0.0
          DO 35 J=0,100
              X = J/100.D0
              ERR = ABS(USOL(0,X)/USOL(0,1.D0) - EXACT(X)/EXACT(1.D0))
              ERMAX = MAX(ERMAX,ERR)
    35    CONTINUE
          PRINT 40, ITER,EIGEN,ERMAX
    40    FORMAT (I5,2E17.8)
    45 CONTINUE
       STOP
       END
       FUNCTION FUNA(X,U,UX,UXX,PK,PKX,PKXX,PI,PIX,PIXX)
       IMPLICIT DOUBLE PRECISION(A-H,O-Z)
C                              INTEGRAND FOR (K,I)-TH ELEMENT OF
C                              'A' MATRIX
       FUNA = EXP(-2.0*X)*(PKX*PIX-PK*PI)
       RETURN
       END
       FUNCTION FUNB(X,U,UX,UXX,PK,PKX,PKXX,PI,PIX,PIXX)
       IMPLICIT DOUBLE PRECISION(A-H,O-Z)
C                              INTEGRAND FOR (K,I)-TH ELEMENT OF
C                              'B' MATRIX
       FUNB = EXP(-2.0*X)*PK*PI
       RETURN
       END
       SUBROUTINE LOCATE(I,HORS,K)
       IMPLICIT DOUBLE PRECISION(A-H,O-Z)
       CHARACTER*1 HORS
C                              DETERMINE THE TYPE ('H' OR 'S') OF BASIS
C                              FUNCTION I, AND THE POINT (XPTS(K)) ON WHICH
C                              IT IS CENTERED.
       IF (MOD(I,2).EQ.0) HORS = 'H'
       IF (MOD(I,2).EQ.1) HORS = 'S'
       K = I/2
       RETURN
       END
       FUNCTION OMEGA(IDER,X)
       IMPLICIT DOUBLE PRECISION(A-H,O-Z)
C                              EVALUATE OMEGA (IDER=0) OR ITS IDER-TH
C                              DERIVATIVE AT X
       OMEGA = 0.0
       RETURN
       END
```

Figure 5.11.1
FORTRAN77 Program to Solve Eigenvalue Problem

are called by this program and thus must be loaded along with these subprograms.

Now consider the two-dimensional eigenvalue problem

$$\mathbf{V}^T(D(x, y)\mathbf{V}\mathbf{u}) = \lambda u \qquad \text{in } (-1, 1) \times (0, 1) \qquad (5.11.2)$$

with homogeneous boundary conditions as shown in Figure 5.11.2.

The function $D(x, y)$ is equal to 1 when $x < 0$ and 2 when $x > 0$. Thus the eigenvalues represent the resonant frequencies of an elastic membrane that is a composite of two different materials, with different elastic properties, which is fastened to a frame along three edges, and is free along the fourth.

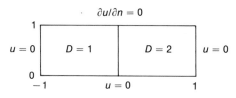

Figure 5.11.2

We again rely on PDE/PROTRAN to solve this two-dimensional problem. Piecewise quartic (4th degree) polynomial basis functions are used in the eigenfunction expansion 5.10.3, so the eigenvalue estimate should have $O(h^8)$ accuracy, where h is a measure of the triangle diameters! The inverse power method 4.11.1 is used to solve the generalized matrix eigenvalue problem $A\mathbf{a} = \lambda B\mathbf{a}$ which results. The contour plot of the first eigenfunction generated by PDE/PROTRAN is given in Figure 5.11.3.

Using only 8 quartic triangular elements (four in each of the two different materials, so that no element straddles the interface) PDE/PROTRAN estimates the first eigenvalue as $\lambda_1 = -6.94234$, which is correct to five significant figures. The material interface at $x = 0$, where $D(x, y)$ is discontinuous, poses no problems for this Galerkin finite element method, and does not even adversely affect the accuracy. This is in marked contrast to the finite difference case, where such an interface must be handled with separate interface conditions $(u(0 - , y) = u(0 + , y)$ and $D(0 - , y)u_x(0 - , y) = D(0 +, y)u_x(0 +, y))$.

An intuitive explanation for the success of the Galerkin finite element method in handling discontinuous material properties is that it uses the weak formulation 5.10.2, in which only integrals of D and the other coefficients (in

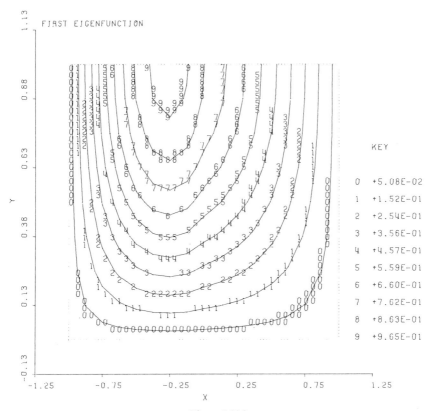

Figure 5.11.3
Contour Plot of First Eigenfunction

undifferentiated form) appear, whereas the finite difference and collocation methods attempt to directly approximate the differential equation, where derivatives of $D(x, y)$ do appear. In addition, since the first derivatives of the eigenfunctions are discontinuous across the interface, it is to be expected that they can be better approximated by the Galerkin trial functions, whose derivatives may be discontinuous across triangle boundaries, than by the collocation trial functions, which have continuous first derivatives.

5.12 Problems

1. a. In the derivation of error bound 5.1.7 it was assumed that the integrals in 5.1.5 are calculated exactly. Assume, on the other hand, that they

are calculated by some numerical quadrature method, and show that 5.1.7 can be replaced by

$$\| U - u \|_F^2 \leqslant \| W - u \|_F^2 + [F_\pi(W) - F(W)] - [F_\pi(U) - F(U)],$$

where F_π is the "energy" functional with its integrals replaced by the corresponding numerical approximations.

Hint: U now minimizes F_π, rather than F, over the set of trial functions.

b. Suppose $u(x)$ is the solution to the one-dimensional version of 5.1.1, and suppose this problem is solved by using a Galerkin method, with cubic Hermite basis functions, and with all integrals evaluated by using a quadrature rule exact for polynomials of degree 5 (e.g., the Gauss 3-point rule). Show that the order of accuracy of $\| U - u \|_F$ is not decreased by the numerical integration.

Hint: Differentiate formula 5.3.8 and then show that $\| W - u \|_F = O(\Delta x_{max}^3)$, where W is the cubic Hermite interpolant to u, and u is assumed to be smooth.

As mentioned in Section 5.3, it is known that a quadrature rule that is exact for polynomials of degree 4 is sufficient to avoid decreasing the order of the Galerkin error, so the above result is not as sharp as it could be made.

2. Determine the finite element equations corresponding to the problem

$$0 = u_{xx} + f(x),$$
$$u(0) = 0,$$
$$u(1) = 0,$$

when the "chapeau" functions of Figure 5.2.1 are used as the basis functions, where the knots x_i are *not necessarily uniformly spaced*, and all integrals are approximated using the trapezoid rule. Since $U(x_k) = a_k$, the resulting tridiagonal linear system (cf. 5.2.2) can be thought of as a finite difference approximation to the differential equation. Show that the truncation error of this finite difference formula is $O(\Delta x_{max}^2)$, as suggested by 5.2.5.

3. Consider the boundary value problem

$$0 = u_{xx} - u + 4\sin(x),$$
$$u_x(0) = 2,$$
$$u(\pi/2) = 2,$$

which has the exact solution $u(x) = 2 \sin(x)$. Solve this problem using a *collocation* finite element method, with cubic Hermite basis functions and collocation points $x_l + \beta_1 \Delta x$, $x_l + \beta_2 \Delta x$, where $\beta_1 = 0.5 - 0.5/\sqrt{3}$ and $\beta_2 = 0.5 + 0.5/\sqrt{3}$. Make runs with $\Delta x = 0.1$ and $\Delta x = 0.05$ and calculate the experimental order of convergence.

Hint: Use the subprograms in Figures 5.4.2 and 5.3.3b as a starting point for your program.

4. Solve the boundary value problem of Problem 3 by using a *Galerkin* method with cubic Hermite basis functions. Evaluate all integrals by using the 3-point Gaussian quadrature rule. Make runs with $\Delta x = 0.1$ and $\Delta x = 0.05$ and calculate the experimental order of convergence. Compare the accuracy and programming effort of the Galerkin and collocation methods on this problem.

Hint: Use formulas 5.1.5, with $D(x) = a(x) = 1$, $f(x) = 4 \sin(x)$, $r = 2$, $p = 0$, and $q = -2$. (Why is q negative?) Then use the subprograms in Figures 5.4.1 and 5.3.3b as a starting point for your program.

5. a. Consider the following fourth-order thin elastic plate problem, in a two-dimensional region R, with both clamped and simply supported boundary conditions:

$$\nabla^2(\nabla^2 u) \equiv u_{xxxx} + 2u_{xxyy} + u_{yyyy} = f(x, y),$$

$$u = r(x, y) \qquad \text{on the entire boundary,}$$

$$\partial u/\partial n = q(x, y) \qquad \text{on part of the boundary } (\partial R_1),$$

$$\nabla^2 u = 0 \qquad \text{on the other part } (\partial R_2).$$

Assume an approximate solution of the form

$$U(x, y) = \Omega(x, y) + \Sigma\, a_i \Phi_i(x, y),$$

where Ω is chosen to satisfy the first two boundary conditions, and the Φ_i are chosen to satisfy $\partial\Phi_i/\partial n = 0$ on ∂R_1, and $\Phi_i = 0$ on all of ∂R. This assures that U satisfies the first two (essential) boundary conditions.

Substitute $U(x, y)$ into the partial differential equation, multiply this equation by Φ_k, and integrate over R. After integrating by parts twice, using the boundary conditions on Φ_k, and assuming $\nabla^2 U = 0$ on ∂R_2, you will obtain the Galerkin method equations for the coefficients a_i in the above expansion for U.

Note that these equations involve integrals of the *second* derivatives of the basis functions. Thus the Ω, Φ_i and their first derivatives must be continuous. This requirement is difficult to satisfy when irregular elements are used, so it is usually preferable to break a fourth-order problem into a system of two second-order equations.

b. Use the equations derived in part (a) to solve the one-dimensional beam problem 4.9.1, using cubic Hermites for the basis functions. (Note that these functions have continuous derivatives, as required.) Choose $\Omega(x) = H_N(x)$, and choose $\{\Phi_1, \ldots, \Phi_{2N-1}\}$ to be the set $\{S_1, H_1, \ldots, S_{N-1}, H_{N-1}, S_N\}$ of cubic Hermite basis functions (see 5.3.5). These choices ensure that $U(0) = U_x(0) = 0$ and $U(1) = 1$. Use the subprograms in Figures 5.3.3 and 5.4.1 as building blocks for your program.

Since the exact solution is a cubic polynomial, your answers should be exact to within roundoff error, even when a small value of N is used (try $N = 4$). Note that collocation using cubic Hermites could not work on this fourth-order problem, since the fourth derivatives of all the basis functions are identically zero.

6. Use IMSL subroutine MOLCH, or a similar method of lines subroutine from your computer library, to solve the time-dependent problem 2.3.1. Notice that MOLCH adaptively varies the time stepsize to control the time discretization error, but the spatial discretizations must be specified by the user.

7. Solve the time-dependent problem 5.8.1 with cubic Hermite basis functions, as outlined in Section 5.8, by using the *first*-order backward difference method. That is, use 5.7.8 with $m = 1$, $\beta_0 = 1$, and $\alpha_1 = -1$, where the elements of A, B, and $\mathbf{b}$ are given in 5.8.2, and the initial values for the $a_i(t)$ are given by equations 5.8.3. Calculate the maximum error at $t = 1$, when $\Delta x = 0.01$, $\Delta t = 0.05$, and compare it with the results using the fourth-order backward difference method, given in Table 5.8.1.

   Hint: Use the programs in Figures 5.3.3 and 5.9.1 as a starting point. Notice that the boundary term in b_k may be omitted, since $\Omega(1, t) \equiv 0$.

8. Calculate the smallest eigenvalue of the two-dimensional eigenvalue problem 5.11.2 by using linear triangular elements to generate the matrix eigenvalue problem $A\mathbf{a} = \lambda B\mathbf{a}$ and the inverse power method 4.11.1 to find the smallest eigenvalue of this discrete problem. Use the triangulation shown below.

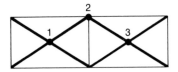

Use Figure 5.9.1 as a starting point for your program. Since $p = 0$, the boundary integral in 5.10.5a is zero, but only those vertices on the portion (∂R_1) of the boundary where $u = 0$ should be assigned the "NODE" number 0.

Notice that the eigenvalue error using linear elements is *very* much greater than the error obtained by PDE/PROTRAN using quartic elements on the same triangulation.

9. Consider the eigenvalue problem

$$u_{xx} + u_{yy} = \lambda u \qquad \text{in } (0, 1) \times (0, 1),$$
$$u = 0 \qquad \text{on the boundary.}$$

Let us use piecewise "bilinear" basis functions and expand our eigenfunction in the form

$$U(x, y) = \sum_{i=1}^{N-1} \sum_{j=1}^{N-1} a_{ij} C_i(x) C_j(y),$$

where the C_i are the piecewise linear "chapeau" functions of Figure 5.2.1, and $x_i = ih$, $y_j = jh$, $h = 1/N$. Clearly, $U = 0$ when $x = 0$, since all the chapeau functions are zero there except $C_0(x)$, which is missing from the expansion. Similarly, $U = 0$ on the other three edges.

This leads to the matrix eigenvalue problem $A\mathbf{a} = \lambda B\mathbf{a}$, where the elements of A and B are given by 5.10.5. It is more convenient here to number the rows and columns of these matrices by using double subscripts, since our basis functions have double-subscript labels. Thus (since $D = 1$, $a = 0$, $\rho = 1$)

$$A_{kl,ij} = \int_0^1 \int_0^1 [-C_k'(x)C_l(y)C_i'(x)C_j(y) - C_k(x)C_l'(y)C_i(x)C_j'(y)] \, dx \, dy$$

$$= -a_{ki}b_{lj} - a_{lj}b_{ki},$$

$$B_{kl,ij} = \int_0^1 \int_0^1 [C_k(x)C_l(y)C_i(x)C_j(y)] \, dx \, dy = b_{ki}b_{lj},$$

where

$$a_{ki} = \int_0^1 C_k'(x)C_i'(x)\, dx = \begin{cases} \dfrac{2}{h}, & \text{if } |k - i| = 0, \\[2mm] -\dfrac{1}{h}, & \text{if } |k - i| = 1, \\[2mm] 0, & \text{otherwise,} \end{cases}$$

$$b_{ki} = \int_0^1 C_k(x)C_i(x)\, dx = \begin{cases} \dfrac{2h}{3}, & \text{if } |k - i| = 0, \\[2mm] \dfrac{h}{6}, & \text{if } |k - i| = 1, \\[2mm] 0, & \text{otherwise.} \end{cases}$$

Use the inverse power method (4.11.1) to find the smallest eigenvalue of the matrix eigenvalue problem. The easiest way to solve the linear system $A\mathbf{u} = B\mathbf{u_n}$ which must be solved for $\mathbf{u}(= \mathbf{u_{n+1}})$ each iteration of the inverse power method, is probably the successive over-relaxation iteration (4.6.3), which, in our double-subscript notation, can be written

$$u_{kl} = u_{kl} - \frac{\omega \left[\displaystyle\sum_{i=k-1}^{k+1} \sum_{j=l-1}^{l+1} [A_{kl,ij}u_{ij} - B_{kl,ij}u_{ij}^n] \right]}{A_{kl,kl}},$$

where the latest calculated values of the components of u are used on the right-hand side. The u_{ij}^n are the components of $\mathbf{u_n}$ and do not change during the successive over-relaxation iteration. It is highly recommended that you store $\mathbf{u}$ and $\mathbf{u_n}$ in double subscripted arrays, along with their boundary values, as was done in Figure 4.7.1.

Set $h = 0.05$ and do 10 iterations of the inverse power method, iterating the successive over-relaxation method to convergence during each inverse power iteration. Estimate the eigenvalue by calculating the ratio of the components of $\mathbf{u_n}$ to the corresponding components of $\mathbf{u_{n+1}}$. The exact value of the first eigenvalue is $-2\pi^2$.

Appendix 1
The Fourier Stability Method

A very useful technique for analyzing the stability of a finite difference method used to solve an initial value problem is the Fourier method, which we have already encountered once, in Section 3.3. Its general application will be illustrated using the diffusion-convection equation 2.1.1.

In all the rigorous stability analyses in Chapters 2 and 3, it was noticed that neither the boundary conditions nor the lower-order terms had any effect on the stability criteria. Thus we discard the lower-order derivative terms and for convenience assume periodic boundary conditions on the interval $[0, 2\pi]$:

$$u_t = Du_{xx} \qquad (D > 0),$$

$$u(0, t) = u(2\pi, t),$$

$$u_x(0, t) = u_x(2\pi, t), \qquad \text{(A1.1)}$$

$$u(x, 0) = h(x).$$

Because it satisfies periodic boundary conditions, u may be extended periodically outside $[0, 2\pi]$ in a smooth manner by $u(x \pm 2\pi, t) = u(x, t)$.

The Fourier method requires constant coefficients, but if D is variable, the finite difference method may be assumed stable if the constant coefficient problem is stable for every constant value of D in its range.

Consider the explicit method (cf. 2.1.3) applied to A1.1:

$$\frac{U(x_i, t_{k+1}) - U(x_i, t_k)}{\Delta t} = D \frac{U(x_{i+1}, t_k) - 2U(x_i, t_k) + U(x_{i-1}, t_k)}{\Delta x^2},$$

$$i = 1, \ldots, N,$$

where $x_i = i \Delta x$ $(\Delta x = 2\pi/N)$ and $t_k = k \Delta t$. U is required to satisfy $U(x_{i+N}, t_k) = U(x_i, t_k)$ for $i = 0, 1$, so it may also be extended periodically outside $[0, 2\pi]$.

The solution u of A1.1 satisfies, by the definition of truncation error,

$$\frac{u(x_i, t_{k+1}) - u(x_i, t_k)}{\Delta t} = D \frac{u(x_{i+1}, t_k) - 2u(x_i, t_k) + u(x_{i-1}, t_k)}{\Delta x^2} + T_i^k,$$

where $T_i^k = O(\Delta x^2) + O(\Delta t)$ is the truncation error. Then the error $e \equiv U - u$ satisfies

$$\frac{e(x_i, t_{k+1}) - e(x_i, t_k)}{\Delta t} = D \frac{e(x_{i+1}, t_k) - 2e(x_i, t_k) + e(x_{i-1}, t_k)}{\Delta x^2} - T_i^k. \quad \text{(A1.2)}$$

Since both U and u are equal to $h(x_i)$ for $t = t_0$, $e(x_i, t_0) = 0$, and since both are periodically extended, $e(x_{i \pm N}, t_k) = e(x_i, t_k)$.

The Fourier method consists in explicitly solving the difference equation A1.2 for the error! Since the N vectors

$$(\exp(Imx_1), \exp(Imx_2), \ldots, \exp(Imx_N)), \qquad m = 1, \ldots, N,$$

$(I = \sqrt{-1})$ can be shown to be mutually orthogonal and therefore linearly independent, they form a basis for the N-dimensional space to which they, and the error at any fixed time, belong. Thus the error at t_k can be expanded as a linear combination of these basis vectors:

$$e(x_i, t_k) = \sum_{m=1}^{N} a_m(t_k) \exp(Imx_i).$$

Note that e is periodic, as required, regardless of the values of the coefficients.

The truncation error is also expanded in this basis:

$$T_i^k = \sum_{m=1}^{N} b_m(k) \exp(Imx_i).$$

If these expansions are plugged into A1.2, we get

$$\sum_{m=1}^{N} \frac{a_m(t_{k+1}) - a_m(t_k)}{\Delta t} \exp(Imx_i)$$

$$= - \sum_{m=1}^{N} b_m(k) \exp(Imx_i)$$

$$+ \sum_{m=1}^{N} D \frac{a_m(t_k)}{\Delta x^2} [\exp\{Im(x_i + \Delta x)\} - 2 \exp\{Im(x_i)\} + \exp\{Im(x_i - \Delta x)\}].$$

Equating the coefficients of $\exp(Imx_i)$ gives

$$\frac{a_m(t_{k+1}) - a_m(t_k)}{\Delta t} = D \frac{a_m(t_k)}{\Delta x^2} [\exp\{Im \, \Delta x\} - 2 + \exp\{-Im \, \Delta x\}] - b_m(k),$$

or

$$a_m(t_{k+1}) = a_m(t_k)\left[1 + D \frac{\Delta t}{\Delta x^2}[-2 + 2\cos(m \, \Delta x)] \right] - \Delta t \, b_m(k),$$

or

$$a_m(t_{k+1}) + \left[-1 + 4D \frac{\Delta t}{\Delta x^2} \sin^2(\tfrac{1}{2}m \, \Delta x) \right] a_m(t_k) = -\Delta t \, b_m(k).$$

By Theorem 1.3.1 we have

$$|a_m(t_k)| \le M_\rho\left[|a_m(t_0)| + t_k \max_{j \le k} |b_m(j)| \right],$$

provided the root of the characteristic polynomial

$$\lambda + \left[-1 + 4D \frac{\Delta t}{\Delta x^2} \sin^2(\tfrac{1}{2}m \, \Delta x) \right]$$

is less than or equal to 1 in absolute value. Since $a_m(t_0) = 0$ (recall that the error is zero at $t = t_0$), the coefficients of the error are bounded by the coefficients of the truncation error. Thus the explicit method is stable, provided

$$|\lambda| = \left|1 - 4D\,\frac{\Delta t}{\Delta x^2}\,\sin^2(\tfrac{1}{2}m\,\Delta x)\right| \leq 1$$

for $m = 1, \ldots, N$. This is true for all m if (recall that $\Delta x = 2\pi/N$):

$$1 - 4D(\Delta t/\Delta x^2) \geq -1$$

or

$$\Delta t \leq \tfrac{1}{2}\,\Delta x^2/D.$$

This stability condition is the same as that given by formula 2.1.7, in the case that D is constant.

In practice, applying the Fourier method is not nearly so complicated as the above analysis might suggest. We merely need to insert an arbitrary term, $a_m(t_k)\exp(Imx_i)$ of the error expansion into the *original* (homogeneous) finite difference equation. (This is the same as the equation satisfied by the error, except for the nonhomogeneous truncation error term, which does not affect the characteristic polynomial.) We then obtain a difference equation for $a_m(t_k)$, and the roots of the characteristic polynomial corresponding to this difference equation are calculated. For stability it is required that these roots, for each m, have absolute value less than or equal to 1 (less than 1, for multiple roots).

For example, if we want to test the stability of the implicit method

$$\frac{U(x_i, t_{k+1}) - U(x_i, t_k)}{\Delta t}$$

$$= D\,\frac{U(x_{i+1}, t_{k+1}) - 2U(x_i, t_{k+1}) + U(x_{i-1}, t_{k+1})}{\Delta x^2} \tag{A1.3}$$

applied to A1.1, we simply insert

$$U(x_i, t_k) = a_m(t_k)\exp(Imx_i)$$

into A1.3, obtaining

$$a_m(t_{k+1}) \left[1 + 4D \frac{\Delta t}{\Delta x^2} \sin^2(\tfrac{1}{2} m \, \Delta x) \right] - a_m(t_k) = 0.$$

The corresponding characteristic polynomial is

$$\left[1 + 4D \frac{\Delta t}{\Delta x^2} \sin^2(\tfrac{1}{2} m \, \Delta x) \right] \lambda - 1.$$

The root of this linear polynomial is always less than 1 in absolute value, so the implicit method A1.3 is always stable, as discovered in Section 2.2.

Table A1.1 shows the results of applying the Fourier method to analyze the stability of various finite difference methods, including some of those studied in Chapters 2–3.

The abbreviations

$$U^k \equiv U(x_i, t_k),$$

$$\Delta_{xx} U^k \equiv \frac{U(x_{i+1}, t_k) - 2U(x_i, t_k) + U(x_{i-1}, t_k)}{\Delta x^2}$$

are used. For those problems in two space dimensions, the Fourier series expansion for the error has terms of the form

$$a_{lm}(t_k) \exp(Ilx_i + Imy_j),$$

and the abbreviations used are

$$U^k \equiv U(x_i, y_j, t_k),$$

$$\Delta_{xx} U^k \equiv \frac{U(x_{i+1}, y_j, t_k) - 2U(x_i, y_j, t_k) + U(x_{i-1}, y_j, t_k)}{\Delta x^2}$$

$$\Delta_{yy} U^k \equiv \frac{U(x_i, y_{j+1}, t_k) - 2U(x_i, y_j, t_k) + U(x_i, y_{j-1}, t_k)}{\Delta y^2}.$$

The constants D, v, c are assumed to be positive.

Table A1.1
Fourier Stability Analysis of Finite Difference Methods

PDE	Method name	Stencil	Formula
$u_t = Du_{xx}$	explicit		$\dfrac{U^{k+1} - U^k}{\Delta t} = D\,\Delta_{xx}U^k$
	implicit		$\dfrac{U^{k+1} - U^k}{\Delta t} = D\,\Delta_{xx}U^{k+1}$
	Crank-Nicolson		$\dfrac{U^{k+1} - U^k}{\Delta t} = D\,\Delta_{xx}[\tfrac{1}{2}U^k + \tfrac{1}{2}U^{k+1}]$
$u_t = D[u_{xx} + u_{yy}]$	explicit		$\dfrac{U^{k+1} - U^k}{\Delta t} = D[\Delta_{xx}U^k + \Delta_{yy}U^k]$
$u_t = -vu_x$	upwind explicit		$\dfrac{U^{k+1} - U^k}{\Delta t}$ $= -v\,\dfrac{U(x_i, t_k) - U(x_{i-1}, t_k)}{\Delta x}$
	centered explicit		$\dfrac{U^{k+1} - U^k}{\Delta t}$ $= -v\,\dfrac{U(x_{i+1}, t_k) - U(x_{i-1}, t_k)}{2\,\Delta x}$
	upwind implicit		$\dfrac{U^{k+1} - U^k}{\Delta t}$ $= -v\,\dfrac{U(x_i, t_{k+1}) - U(x_{i-1}, t_{k+1})}{\Delta x}$
$u_{tt} = c^2 u_{xx}$	explicit		$\dfrac{U^{k+1} - 2U^k + U^{k-1}}{\Delta t^2} = c^2\,\Delta_{xx}U^k$
	implicit		$\dfrac{U^{k+1} - 2U^k + U^{k-1}}{\Delta t^2}$ $= c^2\Delta_{xx}[\tfrac{1}{4}U^{k+1} + \tfrac{1}{2}U^k + \tfrac{1}{4}U^{k-1}]$
$u_{tt} = c^2[u_{xx} + u_{yy}]$	explicit		$\dfrac{U^{k+1} - 2U^k + U^{k-1}}{\Delta t^2}$ $= c^2[\Delta_{xx}U^k + \Delta_{yy}U^k]$

Characteristic polynomial	Stability criterion
$\lambda + \left[-1 + 4D\,\dfrac{\Delta t}{\Delta x^2}\,\sin^2(\tfrac{1}{2}m\,\Delta x) \right]$	$\Delta t \leq \tfrac{1}{2}\,\dfrac{\Delta x^2}{D}$
$\left[1 + 4D\,\dfrac{\Delta t}{\Delta x^2}\,\sin^2(\tfrac{1}{2}m\,\Delta x) \right]\lambda - 1$	always stable
$\left[1 + 2D\,\dfrac{\Delta t}{\Delta x^2}\,\sin^2(\tfrac{1}{2}m\,\Delta x) \right]\lambda + \left[-1 + 2D\,\dfrac{\Delta t}{\Delta x^2}\,\sin^2(\tfrac{1}{2}m\,\Delta x) \right]$	always stable
$\lambda + \left[-1 + 4D\,\Delta t\left\{ \dfrac{\sin^2(\tfrac{1}{2}l\,\Delta x)}{\Delta x^2} + \dfrac{\sin^2(\tfrac{1}{2}m\,\Delta y)}{\Delta y^2} \right\} \right]$	$\Delta t \leq \dfrac{\tfrac{1}{2}\,\Delta x^2\,\Delta y^2}{D[\Delta x^2 + \Delta y^2]}$
$\lambda + \left[-1 + v\,\dfrac{\Delta t}{\Delta x}\,\{1 - \exp(-Im\,\Delta x)\} \right]$	$\Delta t \leq \dfrac{\Delta x}{v}$
$\lambda + \left[-1 + v\,\dfrac{\Delta t}{\Delta x}\,\sin(m\,\Delta x)I \right]$	always unstable
$\left[1 + v\,\dfrac{\Delta t}{\Delta x}\,\{1 - \exp(-Im\,\Delta x)\} \right]\lambda - 1$	always stable
$\lambda^2 + \left[-2 + 4c^2\,\dfrac{\Delta t^2}{\Delta x^2}\,\sin^2(\tfrac{1}{2}m\,\Delta x) \right]\lambda + 1$	$\Delta t \leq \dfrac{\Delta x}{c}$
$\left[1 + c^2\,\dfrac{\Delta t^2}{\Delta x^2}\,\sin^2(\tfrac{1}{2}m\,\Delta x) \right]\lambda^2 + \left[-2 + 2c^2\,\dfrac{\Delta t^2}{\Delta x^2}\,\sin^2(\tfrac{1}{2}m\,\Delta x) \right]\lambda$ $+ \left[1 + c^2\,\dfrac{\Delta t^2}{\Delta x^2}\,\sin^2(\tfrac{1}{2}m\,\Delta x) \right]$	always stable
$\lambda^2 + \left[-2 + 4c^2\,\Delta t^2\left\{ \dfrac{\sin^2(\tfrac{1}{2}l\,\Delta x)}{\Delta x^2} + \dfrac{\sin^2(\tfrac{1}{2}m\,\Delta y)}{\Delta y^2} \right\} \right]\lambda + 1$	$\Delta t \leq \dfrac{\Delta x\,\Delta y}{c(\Delta x^2 + \Delta y^2)^{1/2}}$

Appendix 2
Parallel Algorithms

All comparisons of competing algorithms in this book have been done assuming that they are to be executed on a serial, or "scalar," computer, so that the execution time is simply determined by the total number of operations. A number of computers are now in operation, however, which are able to perform several computations simultaneously. For example, if such a "vector" computer has P processing units, it may be able (neglecting any extra overhead) to add two vectors of length P together in the time it would require a comparable scalar computer to add two scalars together. If the speed-up factor is defined to be the time using one processor divided by the time using P processors, then the speed-up ratio for this problem is close to the optimal value P.

Since it is probable that vector computers will become more and more widely distributed in the future, we are forced to re-examine all our numerical algorithms to see how they will perform in such an environment. Some algorithms may require slight modifications to take advantage of parallel processing abilities, whereas others may have to be completely overhauled.

There are two considerations in writing code for parallel computers. First, the algorithm itself should, to the extent possible, involve calculations that can be done simultaneously. For example, in the recurrence $x_i = rx_{i-1}$, which is expressed in FORTRAN as

```
DO 1 I=2,N
1 X(I) = R*X(I-1),
```

the multiplications must be done sequentially, because $X(3)$ must be calculated before $X(3)$ can be computed, and so forth. On the other hand, the vector assignment $\mathbf{x} = r\mathbf{x}$, which is expressed as

```
DO 2 I=1,N
2 X(I) = R*X(I)
```

is ideally suited for parallel computation, because the N multiplications are mutually independent and can be done simultaneously.

The second consideration, however, is that the algorithm must be coded in such a way that it is apparent to the computer that the calculations can be done simultaneously. In the FORTRAN loop

```
INC = Ø
DO 3 I=1,N
3 X(I) = R*X(I+INC),
```

it is obvious to our eyes that this is equivalent to loop 2, and not loop 1, but the compiler cannot know at compile time what the value of INC will be at execution time, and so it will probably play it safe and generate sequential object code. (Actually, the CRAY-1 compiler will generate both sequential and vector code, and insert a run-time test on INC to determine which code to execute!)

With these ideas in mind, let us look at some of the algorithms presented in the text. First, Gaussian elimination applied to a full matrix, as coded in Figure 0.1.2, vectorizes quite well. The innermost loop (DO loop 30), where for large N most of the execution time is spent, involves the addition of a multiple of row K to row I (to zero element $A(I, K)$), and this operation can be performed on all components simultaneously. A compiler such as the CRAY-1 FORTRAN compiler is able to automatically generate vector code for loop 30, as written in Figure 0.1.2. In fact, if N^2 processors are available, all of the subdiagonal elements in column K can, in theory, be eliminated simultaneously, and the speed-up factor is $O(N^2)$. However, for problems of the size in which we are interested, N^2 is much larger than the number P of processors available, and generally not even one row can be processed in parallel, so the speed-up factor will be approximately P.

Similar comments apply to the band solver (Figure 0.4.3). When the bandwidth is very small, such as for the tridiagonal solver, there is virtually nothing to be gained from parallel processing, but the total work is only $O(N)$ on a scalar computer anyway.

The most time-consuming operation for the sparse direct solvers of Section 0.5 is the addition of a multiple of one row to another. Because of the irregularity of the fill-in pattern, however, only vector computers with "gather/scatter" hardware can perform this operation in vector mode.

None of the algorithms for initial value ordinary differential equations can benefit much from multiple processors, if a single equation is solved, since the solution at t_k must be known *before* the solution at t_{k+1} can be calculated. If a large system of initial value equations is solved, on the other hand, there is potentially some benefit, since on each time step the solution components may be calculated independently of each other. For example, the programs of Figures 2.5.1 and 3.4.2 solve initial/boundary value problems as systems of ordinary differential equations, and the spatial calculations at each time step (loops 10 and 15, respectively) vectorize very well. As a general rule, the solution values at different time points must be calculated sequentially, but the solution values at different spatial points may often be calculated simultaneously.

The conjugate gradient iterative method, used in Figure 4.8.1 to solve the linear equations generated by the finite difference discretization of a steady-state problem, is an example of an algorithm well suited to parallel computation. The dominant calculation is the matrix-vector multiplication $A\mathbf{p}$ which is done in such a way (DO loop 20) that the components of $A\mathbf{p}$ (AP) can be calculated simultaneously. Most of the remaining time is spent adding a multiple of one array to another (subroutine UPDATE) and calculating dot products (function DOT). The dot product operation, as usually programmed, at first does not appear to be well suited to vectorization, since the sum is incremented step by step. However, Figure A2.1 shows how the dot product of two vectors $\mathbf{x}$ and $\mathbf{y}$ *can be* calculated efficiently when multiple processors are available. It is easy to see that if the lengths of $\mathbf{x}$ and $\mathbf{y}$ are N, the total number of levels in the pyramid is about $\log_2 N$, and thus if a sufficient number of processors are available, the total run time will be $O(\log N)$, compared to $O(N)$ on a scalar computer.

On a CRAY-1, at least, it is not necessary to explicitly order the dot product calculations as indicated by Figure A2.1. The CRAY FORTRAN compiler is able to take the function DOT as written in Figure 4.8.1 and to produce efficient vector code, although the calculations are not done in exactly the order shown in Figure A2.1.

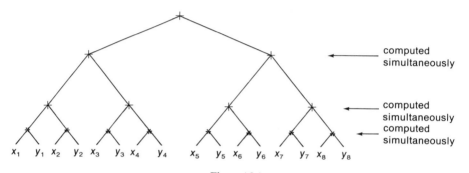

Figure A2.1
Parallel Computation of a Dot Product

When the conjugate gradient program of Figure 4.8.1 is run on the CRAY-1 with the compiler vectorization switch alternately off and on, a speed-up factor of about 3.68 is observed experimentally, using single precision. The CRAY-1 is really a "pipeline" computer, which means that the components of a vector are not processed simultaneously by several processors, but instead are operated on in "assembly line" fashion by a single processor, which begins working on the second component before it has finished with the first. Nevertheless, the CRAY-1 exhibits a maximum speed-up factor of about 6–8 so that we can consider it, for our purposes, as an array processor with 6–8 processors. If we use the proximity of the observed speed-up factor to the maximum possible as a measure of the suitability of an algorithm for parallel computation, then the program of Figure 4.8.1 receives a fair grade. When N is greater than 40, the speed-up factor increases further, as overhead becomes less significant.

The successive over-relaxation algorithm, on the other hand, is very poorly suited for vectorization, as coded in Figure 4.7.1. On each iteration, the new components of the solution array must be calculated sequentially, according to 4.6.3. The experimentally observed speed-up for the program of Figure 4.7.1, as written, is only 1.01, even when the error calculation is removed from DO loop 15.

However, upon rethinking the successive over-relaxation algorithm, we see that it can be made more suitable for parallel calculations by changing the order in which the solution components are updated. Instead of ordering the unknowns (and associated equations) in "typewriter" order, let us use a "red-black" ordering. That is, the unknowns $U(x_i, y_j)$ with $i + j$ even (corresponding to the red squares on a chess board) are updated before any of the unknowns with $i + j$ odd (the black squares). Since the formula used to update each red unknown involves no other red unknowns, all the red unknowns may be updated simultaneously, and then the black unknowns may be updated simultaneously.

This updating order is accomplished by replacing the main loop (DO loop 15) in Figure 4.7.1 by

```
C                          UPDATE RED UNKNOWNS
      DO 15 I=1,N-1,2
      DO 15 J=1,N-1,2
         UR(I,J) = UR(I,J) - W*( (-UB(I+1,J)+2*UR(I,J)-UB(I-1,J))/DX**2
     &                          +(-UB(I,J+1)+2*UR(I,J)-UB(I,J-1))/DY**2
     &     + A*UR(I,J) - X(I)*Y(J)*(A*Y(J)**2-6) ) / (4/DX**2+A)
   15 CONTINUE
      DO 16 I=2,N-1,2
      DO 16 J=2,N-1,2
         UR(I,J) = UR(I,J) - W*( (-UB(I+1,J)+2*UR(I,J)-UB(I-1,J))/DX**2
     &                          +(-UB(I,J+1)+2*UR(I,J)-UB(I,J-1))/DY**2
     &     + A*UR(I,J) - X(I)*Y(J)*(A*Y(J)**2-6) ) / (4/DX**2+A)
   16 CONTINUE
C                          UPDATE BLACK UNKNOWNS
      DO 17 I=1,N-1,2
      DO 17 J=2,N-1,2
         UB(I,J) = UB(I,J) - W*( (-UR(I+1,J)+2*UB(I,J)-UR(I-1,J))/DX**2
     &                          +(-UR(I,J+1)+2*UB(I,J)-UR(I,J-1))/DY**2
     &     + A*UB(I,J) - X(I)*Y(J)*(A*Y(J)**2-6) ) / (4/DX**2+A)
   17 CONTINUE
      DO 18 I=2,N-1,2
      DO 18 J=1,N-1,2
         UB(I,J) = UB(I,J) - W*( (-UR(I+1,J)+2*UB(I,J)-UR(I-1,J))/DX**2
     &                          +(-UR(I,J+1)+2*UB(I,J)-UR(I,J-1))/DY**2
     &     + A*UB(I,J) - X(I)*Y(J)*(A*Y(J)**2-6) ) / (4/DX**2+A)
   18 CONTINUE
```

The arrays UR and UB hold the values of U at the red and black points, respectively. Before the iteration begins, both are initialized like U in Figure 4.7.1.

References

Duff, I., A. Erisman, and J. Reid (1987), *Direct Methods for Sparse Matrices*, Oxford University Press.

Forsythe, G. and W. Wasow (1960), *Finite Difference Methods for Partial Differential Equations*, John Wiley and Sons.

Grimes, R., D. Kincaid, W. MacGregor, and D. Young (1978), "ITPACK Report: Adaptive Iterative Algorithms Using Symmetric Sparse Storage," CNA-139, University of Texas.

Hageman, L. and D. Young (1981), *Applied Iterative Methods*, Academic Press.

Hindmarsh, A. (1974), "GEAR: Ordinary Differential Equation System Solver," Laurence Livermore Laboratory Report UCID-30001.

Hindmarsh, A. (1984), "ODE Solvers for Time-Dependent PDE Software," p. 325 in *PDE SOFTWARE: Modules, Interfaces and Systems*, North-Holland.

Horn, R. and C. Johnson (1985), *Matrix Analysis*, Cambridge University Press.

Mitchell, A. and D. Griffiths (1980), *The Finite Difference Method in Partial Differential Equations*, John Wiley and Sons.

Pereyra, V. (1978) "PASVA3: An Adaptive Finite Difference FORTRAN Program for First-Order Nonlinear Ordinary Boundary Problems," Lecture Notes in Computer Science **76**: 67–88, Springer-Verlag.

Rice, J. and R. Boisvert (1985), *Solving Elliptic Problems Using ELLPACK*, Springer-Verlag.

Sewell, G. (1982), "IMSL Software for Differential Equations in One Space Variable," IMSL Technical Report 8202.

Sewell, G. (1985), *Analysis of a Finite Element Method: PDE/PROTRAN*, Springer-Verlag.

Sewell, G. (1988), "Plotting Contour Surfaces of a Function of Three Variables," ACM Transactions on Mathematical Software, March 1988.

Smith, B., J. Boyle, B. Garbow, Y. Ikebe, V. Klema, and C. Moler (1974), *Matrix Eigensystem Routines—EISPACK Guide*, Springer-Verlag.

Strang, G. and G. Fix (1973), *An Analysis of the Finite Element Method*, Prentice-Hall.

Varga, R. (1971), *Functional Analysis and Approximation Theory in Numerical Analysis*, SIAM Regional Conference Series in Applied Mathematics.

Index